AIN'T THERE NO MORE

AIN'T THERE

Carl A. Brasseaux and Donald W. Davis

Foreword by Robert Twilley

Carl A. Brasseaux and Donald W. Davis, series editors

University Press of Mississippi / Jackson

NO MORE

Louisiana's Disappearing Coastal Plain

This contribution has been supported with funding provided by the Louisiana Sea Grant College Program (LSG) under NOAA Award # NA14OAR4170099. Additional support is from the Louisiana Sea Grant Foundation. The funding support of LSG and NOAA is gratefully acknowledged, along with the matching support by LSU. Logo created by Louisiana Sea Grant College Program.

This project was funded in part by a donation to Louisiana Sea Grant Foundation from the estate of Dr. Jack Van Lopik, in honor of the services to Louisiana Sea Grant by Ronald E. Becker.

www.upress.state.ms.us

The University Press of Mississippi is a member of the Association of American University Presses.

Designed by Todd Lape

Manufactured in Malaysia

First printing 2017
∞

Library of Congress Cataloging-in-Publication Data

Names: Brasseaux, Carl A., author. | Davis, Donald W. (Donald Wayne), 1943– author.
Title: Ain't there no more : Louisiana's disappearing coastal plain / Carl A. Brasseaux and Donald W. Davis ; foreword by Robert Twilley.
Description: Jackson : University Press of Mississippi, [2017] | Includes bibliographical references and index.
Identifiers: LCCN 2016029067 (print) | LCCN 2016042987 (ebook) | ISBN 9781496809483 (hardcover : alk. paper) | ISBN 9781496809490 (epub single) | ISBN 9781496809506 (epub institutional) | ISBN 9781496809513 (pdf single) | ISBN 9781496809520 (pdf institutional)
Subjects: LCSH: Gulf Coast (La.)—Environmental conditions. | Gulf Coast (La.)—Population. | Coast changes—Louisiana. | Coastal plains—Louisiana.
Classification: LCC GE155.L8 B73 2017 (print) | LCC GE155.L8 (ebook) | DDC 333.91/71609763—dc23
LC record available at https://lccn.loc.gov/2016029067

British Library Cataloging-in-Publication Data available

To Harley Jesse "Jess" Walker, Boyd Professor,

Louisiana State University, 1921–2015,

and A. David Barry, Professor and Dean Emeritus,

University of Louisiana at Lafayette

FOREWORD

A well-informed public is more likely to engage constructively in science-related policy decisions. Government policies, management decisions, and corporate and personal decisions should be informed by the best scientific knowledge available.

Louisiana Sea Grant's mission, as clearly articulated in the policy statement set out to the right, is to promote stewardship through a combination of programs critical to the cultural, economic, and environmental health of the Bayou State's coastal zone. In Louisiana, one of only two states actually encompassing portions of the Mississippi River—the other being Minnesota—there is a very strong connection between the coastal zone and the Mississippi River delta. The Mississippi River, a drainage basin that connects some 41 percent of the coterminous United States, is the fourth largest riverine system in the world in terms of drainage area and the seventh largest in terms of discharge and sediment load. As such, the Mississippi River basin's watershed integrates the effects of human-engineered river basin processes that forge linkages among economic, environmental, and social outcomes. These outcomes, in turn, provide insights necessary to build a sustainable future by working with, not against, nature. But efforts to construct that future must take stock regarding how changes in the social fabric of a deltaic landscape are linked to engineering decisions regarding river management.

Much of the available information regarding that linkage has been compiled in recent years by Louisiana Sea Grant scholars, who have assembled a large and continuously growing digital collection of traditionally overlooked documents, historic images, material culture, industrial memorabilia, and oral histories. The oral histories are particularly significant, because they collectively constitute the coastal communities' environmental knowledge accumulated over more than two centuries.

Books in the Third Coast series, which mine and interpret these underutilized resources, constitute pioneer efforts to provide the general public with authoritative guides to how human settlement and river management decisions have molded the way we, as a nation, value ecosystems along a major alluvial valley. The history of changes in human settlement and ecosystem services in the Mississippi River delta chronicles the best and the worst attempts to balance economic development, national priorities in navigation and agriculture, and public safety. The consequences include several environmental catastrophes at the coastal end of the river basin, where the highest wetland loss rate and largest hypoxic zone in North America exists. Now, there is renewed effort to restore the Mississippi River delta's wetlands and reduce the runoff of excessive nutrients into the Gulf of Mexico. Such bold actions within the river basin are complicated and limited by staggering costs as well as federal mandates to control river floods, maintain navigation, and promote agricultural and industrial production.

The origins, evolution, and consequences of this state of affairs are explored in *Ain't There No More*, the first number of the Third Coast series. The authors, both Louisiana Sea Grant scholars, explore what we've lost as a people while engineering a river basin that has created tremendous wealth but also ecosystem damage to its downstream delta. Flood control on the Mississippi River has interrupted the river's natural deltaic cycle, preventing the delivery of sediment that maintains the land-building processes of the delta region.

One of the most critical benchmarks in evaluating the co-evolution of natural processes of a delta and human settlement can be found in "The Delta of the Mississippi River," *National Geographic Magazine* (December 1897). The author, E. L. Corthell, clearly delineates some of the fundamental bargains forged between humans and nature:

> No doubt the great benefit to the present and two or three following generations accruing from a complete system of absolutely protective levees, excluding the flood waters entirely from the great areas of the lower delta country, far outweighs the disadvantages to future generations from the subsidence of the Gulf delta lands below the level of the sea and their gradual abandonment due to this cause.

Nearly five generations after *National Geographic*'s 1897 benchmark report, Mississippi Delta residents struggle to develop restoration plans within the constraints imposed by federal legislation governing Mississippi River flood control. Affected residents and decision makers face the seemingly insurmountable challenge of compiling, thematically organizing, and systematically interpreting huge, complex, and continually expanding institutional and traditional knowledge bases. Access to dependable data analyses is particularly essential to communities impacted by environmental policy decisions to assess accurately the consequences, particularly the tradeoffs inevitably necessary to sustain human occupation of the coastal plain. The Third Coast book series, which provides readily accessible interpretations of the burgeoning knowledge base by independent experts, will be a critical component in that deliberative process.

—Robert Twilley, Executive Director
Louisiana Sea Grant College Program

PREFACE

Louisiana's coastal plain "is a unique area of the United States' natural heritage; a pinion of domestic energy security, a vessel of cultural traditions; a clustering of tightly knit extended families, natural resource extraction communities, and a vibrant marketplace serving the entire United States. If policy makers fail to come to grips with these economic and civic realities and thus fail to commit to a long-term significant program of barrier island and shoreline protection, the human landscape of coastal Louisiana will ultimately be represented merely by a random and nostalgic collection of relics, memories, and reminiscences of lost opportunities."

Source: S. Laska, G. Wooddell, R. Hagelman, R. Gramling, and M. T. Farris, "At Risk: The Human Community and Infrastructure Resources of Coastal Louisiana," *Journal of Coastal Research* (2005): 110.

The states constituting America's often overlooked Third Coast find themselves on the edge of unprecedented change as a result of successive natural and man-made disasters—some of biblical proportions—that have individually, and collectively, reshaped the region's face over the past decades at the cost of many lives and immense human suffering. Confronted with severe and extensive coastal erosion, as well as rising ocean levels, temperatures, and acidification—exacerbated by pollution and coastal subsidence—state, federal, corporate, and private agencies have attempted to address these problems through massive expenditures earmarked for engineering projects and scientific (read hydrological and biological) research. Very little of this money is dedicated to the human dimension of the problem. As a consequence, policy makers are formulating policies for communities they have never visited and literally know next to nothing about. However, in the policy makers' defense, relatively little relevant material is readily available in print, and, of that literary corpus, almost nothing is geared to a general readership. This is particularly true of the industries and human activities directly impacted by unfolding environmental change and the aforementioned agencies' remedial responses. For example, there are no accessible comprehensive histories of the Gulf Coast's shrimp, oyster, and trapping industries, as well as the communities involved in their operation and maintenance.

Such works are essential for two reasons: they are absolutely necessary to inform policy makers' decisions that will irrevocably change the lives and destinies of Gulf Coast communities; and they are equally indispensable to members of coastal communities who know painfully little about their own background and how the looming threats came into existence. In both instances, the following age-old proverb applies: "you can't know where you're going if you don't know where you've been."

But we, as a people, can't ascertain where we've been without the vital expertise of individuals who have thoroughly immersed themselves in the region's cultural, economic, and social history—particularly where each topic (subdiscipline) intersects the area's environmental history. Such expertise is, unfortunately, in exceedingly limited supply, further magnifying the importance of the contributions to the University

> Consideration of time changes must inevitably make for a richer and fuller geography. It contributes to a better understanding of the present and an increased ability to project into the future. It curbs a dangerous tendency to find too-ready causal relationships between man and nature, and instills a respect for historical processes.

Source: F. Kniffen, "Geography and the Past," *Journal of Geography* 50(3) (1951): 129.

Press of Mississippi's new Third Coast series. Drawing upon their ninety-plus years of cumulative experience in fieldwork and documentary research in the upper Gulf Coast region, the series editors have collaborated to produce the first number in the series, *Ain't There No More,* which explores what we've lost as a people on our collective journey to the precipice of irrevocable, perhaps calamitous change.

ACKNOWLEDGMENTS

Telling this story with a useful degree of accuracy and understanding has required much help. To all of those who have contributed, directly or indirectly, to the information so essential to this narrative, we say "Thank you." Consequently, this book has been shaped by many hands and resources from an assortment of public and private libraries' collections and archives, along with a mix of paper goods, records of all types, and timely oral history accounts. In some cases, we had to seek out, investigate, and fully digest the oldest extant records from sources best described as at the margins of mainstream research. While looking for forgotten, or ignored, material, we had to coax complementary information from numerous nontraditional sources, such as lithographs, material culture, eighteenth- and nineteenth-century newspapers, canning labels, post cards, and early photography. Every bit of unearthed minutia became tesserae in the complex mosaics that coalesced to form our interpretative visions of the assembled material used in this book.

Without the assistance and encouragement of Chuck Wilson, former director of the Louisiana Sea Grant College Program, and the current director, Robert Twilley, along with all of the Sea Grant family, it would have been difficult to find, assemble, and collate the material used as the backdrop to complete this manuscript in a timely matter. We thank all of them, collectively and individually, for allowing two old warhorses an opportunity to continue the passions we honed in graduate school, while we followed our desires to uncover material overlooked, ignored, or unnoticed by individuals who believed our pursuits were a waste of our time.

Finally, we must thank our spouses, Glenda and Karen. They not only worked while we were in graduate school, but over the years have coped with our absences, our penchant to collect material that appeared to be poor investments, and our willingness, in a minute, to drop our work and run the back roads of Louisiana's coastal lowlands. Both have shared our troubles and frustrations, offered encouragement when it appeared we had "hit the wall," and always smiled as we struggled to learn and produce what appears herein. They knew, from years of dealing with our near eccentric behavior, that our unwavering desire to finish would overcome our short-lived struggles. We love them both.

The Louisiana coastal plain's principal cities, towns, and villages.
(Map compiled and edited by Lisa Pond, cartographer)

Principal waterways in the Louisiana coastal plain.
(Map compiled and edited by Lisa Pond, cartographer)

INTRODUCTION

Since the formation of the Mississippi Delta and its associated coastal wetlands began more than ten millennia ago, South Louisiana's physical and cultural landscapes have been works in progress, and the evolutionary processes molding the coast's complex environments continue unabated to the present. Over the past two centuries, however, the natural forces that hitherto shaped the physical landscape have taken a backseat to the modifications wrought by humans seeking to harness the region's riparian resources for short-term economic gains. Formerly populated areas that once supported oak groves, freshwater cypress swamps, and farms are now open water. In order to understand the underlying motivation for significant human intervention in the wetlands, one must first comprehend regional attitudes regarding the coast, particularly the disconnect between the traditionally irreconcilable respective viewpoints of locals and outsiders regarding the importance of littoral wetlands.

In colonial times, provincial officials expressed interest in the present Louisiana coast only when imperial administrators required periodic assessments of Louisiana's maritime timber and naval stores reserves, sailors were shipwrecked or marooned along remote beaches, or a perceived threat of invasion required a survey and assessment of potential invasion routes. Only one settlement within the present state boundaries was established and maintained within sight of the Gulf of Mexico—and then only because it was absolutely necessary to maintain navigation across the treacherous bar at the mouth of the Mississippi River.

Instead, official policy encouraged settlement and development of lands on the natural levees bordering the region's major waterways, which then functioned as the region's principal communications arteries. Lower Bayou Lafourche, for example, was not officially explored—at the colonial governor's behest—until 1773.

State and national policy makers did little to alter this now deeply rooted coastal policy, even when circumstances conspired to focus public attention on the coastal plain. State officials largely overlooked the steady migration of settlers into the coastal wetlands. Most of these settlers were forced from their families' original farmsteads on the interior natural levees by forced heirship, increasingly stringent levee construction and maintenance regulations, and growing pressure to relocate or adapt to the region's emerging plantation economy.

Migrating to undesirable lands on the margins of the colonial settlements, successive generations of these economic refugees gradually occupied lands close and closer to the Gulf shores. Physically isolated in these new venues, these migrants became, at best, an afterthought for the inland communities and for the region's planter-dominated parish governments. Indeed, their northern neighbors generally took no notice of coastal wetland communities, except in the wake of deadly hurricanes, whenever short-lived coastal resorts were established, or, in the late nineteenth century, when commercial hunting and fishing provided South Louisiana urbanites a steady supply of seafood and wild game. It is thus hardly surprising that most coastal communities had no roadway access to the interior until the post–World War II era—and then only because the focus of oil and gas exploration had shifted from the interior to the marshlands and near coastal waters.

The neglect of the coastal communities by local, state, and national governments was not entirely benign. For example, from the earliest days of exploration and colonization, government-sponsored cartographers—when they attempted to denote coastal plain features at all—routinely identified them with such pejorative labels as "useless lands." This incipient attitude took root among colonial and, later, state officials, who came, over time, to equate the value of the marsh dwellers—who paid minimal property taxes, if any—with their patently "worthless" surroundings. It is thus hardly surprising that officeholders routinely promoted the interests of the propertied inlanders at the expense of the coastal outlanders (a trend that has not appreciably changed to the present).

For centuries, interlopers have openly denigrated the residents of coastal wetlands, who have stubbornly refused to abandon the Bayou Country and the *prairies tremblants* in the wake of repeated natural and man-made disasters. Yet the accumulated environmental knowledge that these wetlands survivors have gathered through painful historical experiences holds invaluable keys to the successful adaptation and survival of modern coastal communities elsewhere, as ongoing climate change forces them to contend with rising sea levels, coastal erosion, and, inevitably, difficult lifestyle choices.

Here, as elsewhere, the most insidious challenges are man-made. Since channelization of the Mississippi River in the wake of the 1927 flood, which diverted sediments and nutrients from the wetlands, coastal Louisiana has lost to erosion, subsidence, and rising sea levels a landmass roughly twice the size of the state of Connecticut.

State and national policy makers blissfully ignored this unfolding environmental Armageddon until Hurricane Katrina focused a harsh spotlight on the human consequences of eight decades of neglect. Yet, even today, the welfare of Louisiana's coastal plain residents remains, at best, an afterthought in state and national policy discussions.

This disturbing indifference is the product of deep-rooted popular attitudes, informed by centuries-old stereotypes. Since the seventeenth century, outside observers have consistently viewed the coastal wetlands as a harsh, dangerous, and all-but-impassible "no man's land." Outsiders likewise have consistently regarded the region's residents as mysterious, exotic, and wretched as the land they occupy. As late as the dawn of the twenty-first century, these attitudes were pervasive throughout the country, according to state-sponsored polling.

These seemingly intractable attitudes—continuously reinforced by a barrage of recent nationally televised "reality" programs—obscure and deride the economic and environmental importance of Louisiana's coastal communities. First, coastal residents provide critical logistical support for the nation's most important offshore oil operations. Simply put, without these folks and their specialized skills, the strategically important offshore oil industry in the Gulf of Mexico simply could not exist. Second, because coastal Louisiana has the nation's most sedentary population, the locals possess more than two centuries of accumulated environmental knowledge regarding the fragile wetlands. They *are* the experts. For more than 200 years they have successfully withstood the challenges now confronted by the post-Sandy survivors on Staten Island and the Jersey Shore. Yet policy makers consistently ignore or belittle such invaluable expertise, effectively viewing coastal residents collectively as an irritating complication to plans formulated hundreds of miles from the coast.

This situation, not surprisingly, has elicited frustration and anger from Louisiana's coastal residents. For coastal families, the current coastal policy dialogue is by no means a scholarly debate about abstract problems. Indeed, in some century-old communities, residents now find Gulf waters literally lapping at their doorsteps. The unfolding emergency—and the increasingly acerbic public indictment of untenable policies—demands new approaches to the human dimension of coastal problems. It is imperative that policy makers think locally and embrace the residents' deep-rooted love of place in current restoration planning and implementation efforts. The time has come to invite the area's inhabitants to join in meaningful discussions to find workable solutions to coastal issues, and "practitioners" of the social sciences must assume roles as ombudsmen in the effort to initiate and sustain a constructive dialogue between the people and policy makers. But such a dialogue will be possible only when the latter realize that the “folks” are key to the resolution of otherwise insoluble coastal problems.

Greater depth of understanding is also required within the coastal communities, for a disconnect also exists between generations all across the region’s enthnoracial spectrum. Members of post–World War II generations, who now often vigorously push back against coastal restoration projects for fear of damage to fisheries, have witnessed unprecedented environmental change, diminished economic expectations, and narrowing employment options during their lifetimes. Increasingly cut off from the life experiences of their elders by linguistic barriers and very different life experiences, they no longer appreciate that healthy, environmentally nurturing freshwater streams and saltwater fisheries once peacefully coexisted throughout the coastal plain. Indeed, the *long-term* prosperity of the seafood industry was—and remains—utterly dependent upon the welfare of critical freshwater resources.

The Third Coast Series of the University Press of Mississippi seeks to bridge these very real, very acrimonious, and very debilitating disconnects that threaten the very existence of a distinctive coastal way of life more than two centuries in the making. Written by academic researchers who have devoted their lives to the study of the Gulf Coast and its inhabitants, the series’ component volumes will explore the major facets of the region’s cultural, physical, and economic landscapes, in the process hopefully making sense of the kaleidoscopic cultural, economic, and physical features that have hitherto defied accurate, comprehensive analysis. It is the editors’ hope that this series will impart authoritative information that will provide the foundational basis for *informed* decisions inside and outside the coastal communities.

AIN'T THERE NO MORE

Dying/dead vegetation provides stark evidence of the ecological changes now occurring in Louisiana's coastal zone. These changes directly threaten the local residents' traditional way of life. (Photo by the authors, 1989)

1

MESSIN' WITH MOTHER NATURE

During the past fifty years great changes have taken place in this State due to man-made factors which are changing the conditions of existence for our wild life and fisheries from its very foundations. Perhaps the most significant of these is flood protection in the valley lands of our rivers by means of levees which render the front lands and considerable areas of the back lands reasonably free from annual floods. . . . The situation has further been exaggerated by the fact that Louisiana has also to bear the brunt of the burden due to flood protection, drainage and deforestation going on in the entire area of 1,325,000 square miles of the Mississippi valley.

Source: P. Viosca, "Louisiana Wet Lands and the Value of Their Wild Life and Fishery Resources," *Ecology* 9(2) (1928): 221–222.

South Louisiana's physical landscape is remarkably young, its history comprising mere milliseconds on a geologic time line. The region's complex mélange of land, mud, and water is a direct product of the notorious volatility, bifurcating tendencies, and sedimentation patterns of the Mississippi River and its associated deltas over the course of recent millennia. In studying the Holocene sediment history of the Mississippi River's drainage basin (the world's fourth largest), coastal geomorphologists have divided Louisiana's organic lowlands into the *Chênière* (often Anglicized to Chenier) and Delta Plains. In southwest Louisiana, each *chênière* (ancient beach ridge with an average elevation of about six feet) is perched on a muddy substratum. These marsh ridges constitute part of an archipelago of inland "islands"—collectively referred to as the *Chênière* Plain—each marking the position of a once-active ancient shoreline. These primeval beaches are a product of the Mississippi's prehistoric meanderings. When the river occupied its westernmost prehistoric course, littoral currents carried clay, mud, and sand westward, advancing the *Chênière* Plain as a mud coast in the Louisiana and Texas lowlands west of Marsh Island. Interruptions in the progradation process allowed coarser particles to accumulate, forming a ridge. The increase in sedimentation also caused the shoreline to advance, leaving behind conspicuous *chênières* as the region's most impressive and distinctive topographic features. These relatively rare natural phenomena—scattered across the southwestern marshes—provide residents of the coastal wetlands with a modicum of "high"-ground protection against hurricanes.

The Delta, or Deltaic Plain, lies east of Marsh Island. This region consists of alluvial material deposited over the past 7,500 years by the Mississippi River as a by-product of its dynamic confrontation with the Gulf of Mexico. The river's headlong collision with the Gulf Stream results in a rapid loss of the freshwater current's velocity and

Wetland loss stems in part from engineering efforts to tame the once unfettered Mississippi in response to settlers' outcries for relief from the very flooding responsible for creation and maintenance of the lands upon which they lived. This system of earthen bulwarks has harnessed and trained the great river's distributaries to follow an unnaturally narrow, designated route to the Gulf.

From the beginning, efforts to channelize the Mississippi River were fraught with problems. Levee systems were originally constructed and maintained by landowners. Following the Louisiana Purchase in 1803, the federal government demonstrated no interest in flood protection; however, Reconstruction-era floods (1862–1877) and unprecedented damage from the inundations of 1882, 1892, and 1927 forced Congress to become proactive in flood management. Beginning in 1928, the national government built—at a cost of $5.8 billion—the present "guide levee" system, which produced short-term economic benefits at the expense of long-term, perhaps irreparable environmental costs. Critical nutrients and sediments no longer reached dependent ecosystems, and saltwater consequently began to intrude into formerly flourishing brackish, intermediate, and freshwater habitats, slowly destroying them.

The disruption of the coastal region's natural hydrological plumbing also compounded the problem of environmental damage wrought by major navigational channels and oil and gas exploration in the twentieth century. In the 1930s the focus of Louisiana's oil industry shifted from the state's northern parishes to the coastal zone, particularly the coastal marshes, where remote drilling sites were typically accessible only to barge-mounted rigs. Between roughly 1930 and the turn of the twenty-first century, oil companies dredged canals to serve the approximately 67,000 completed wells in the coastal wetlands and additional canals to accommodate pipelines transporting oil and gas to processing and distribution facilities in the interior. Canals destroyed the integrity of the coast's *prairies tremblants*, facilitated erosion by allowing tidal action to work on fragile wetland environments, and permitted greater saltwater intrusion to destroy indigenous grasses that were essential to the marshes' well-being. Finally, spoil banks resulting from dredging activity disrupted what little natural hydrology remained.

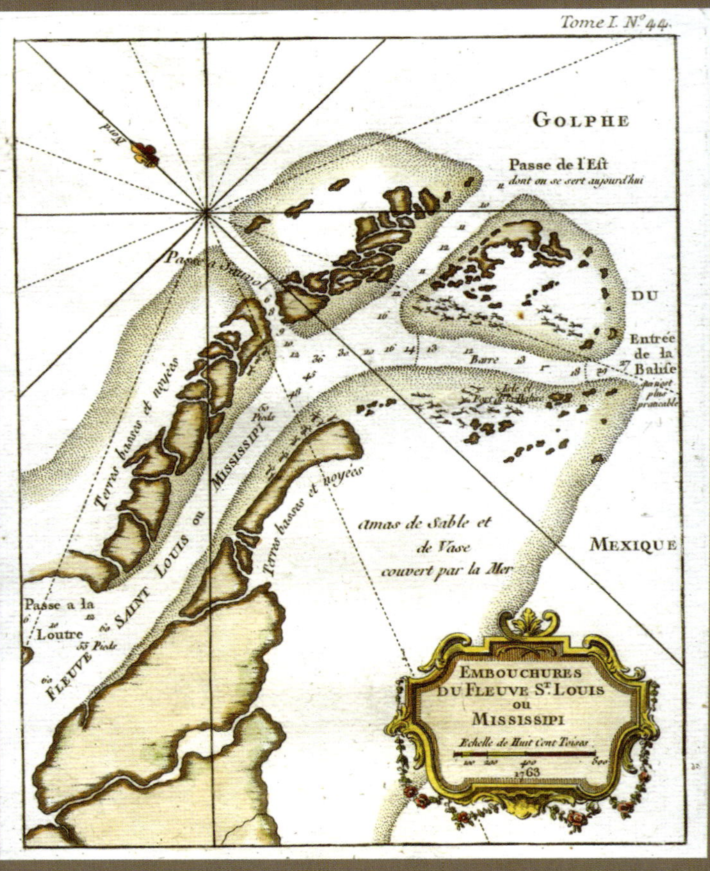

Published in Paris in 1763 by the cartographer J. N. Bellin, this map shows the primary navigation channel, with depth soundings, for waterborne commerce into and out of the Mississippi River. Tiny illustrations on the maap show the accumulation of logs washed down and lodged in the silt deposits of shallow waters, as well as the location of the delta community of Balize. (From the authors' collections)

LATEST FROM ORLEANS.

Copy of a letter from a gentleman in New-Orleans to his friend in this city, dated Feb. 3.

Since my last the enemy has embarked the whole of his army and is now off Ship Island. Our prisoners have been exchanged. Those of our company who had been taken, arrived two days ago. Among them were G. V. Ogden, Pollock, Lynn, Laverty, Doctor Cochrane, &c. &c. Dr. Flood went to the fleet to attend them. They state that the enemy is occupied in building a number of boats, and is believed to be scarce of provisions. We cannot yet ascertain whether or not he intends paying us another vistt. I think it probable that he either expects and is waiting for a supply of provisions to move off, or is waiting for a rupply of provisions to move off, or is waiting a reinforcement. If he could not succeed after landing his whole force, it is not probable that he will again attempt it with a disheartened remnant of a shattered army, and against an augmented and inspirited force. Mr. Ogden says that the enemy landed 15,000 troops, 11,000 of the line and 4,000 marines and sailors. Dr. Flood states, that from what he saw when on board their fleet, he thinks they must have nearly 1500 wounded. He understood they had lost a great number, say 3 or 400, in drowned while passing from and to their shipping; that one of their boats with 80 men sunk, and not a soul saved.

Their loss in various ways has been immense.—They acknowledge they have lost two of their best generals, and that we have beaten the finest troops that Europe can boast of. Gen. Keane is not dead as it was believed when I last wrote. He received two wounds which would have been mortal had not the balls been spent.

The British officers, I understand, bestow great praise on Gen. Jackson. He certainly deserves more than eulogy from his country—His exertions saved this country. I have no donbt but that we shall beat the enemy, let the point or points of attack be made where they may. Mr. Duffy, who arrived to-day, states that he passed, one hundred miles above Natchez, about two thousand volunteer Kentuckians, all armed, coming to join us.—This is the first intimation we have had of this force. If they continue to come down they will really " glut the market." A large force in this country is indispensable.—This state is an import-link in the federate chain, not only as regards the local interests of the country above, but as regards the union of the states. We have been most providentially saved. I am now fully convinced, that had the enemy succeeded (being master of the river, the [illegible]) in getting possession of this country, he could not have been dispossessed again, at least from Baton Rouge down.

The war here has given rise to a great many good anecdotes. The members of our company, while prisoners, being found to be gentlemen of distinction, occasionally dined with the admiral and other of the officers of rank. The Admiral observed that he was astonished to find such men as they were, and merchants too, bearing arms against the British, who came to give them commerce—that they expected, when they came, to find balls and suppers given. Laverty answered, " we have given you the *balls*, you must now look out for the suppers." The admiral, turning to one of the officers present, said, " take that out of your wig."

The Mississippi River Basin. (Map compiled and edited by Lisa Pond, cartographer)

The Battle of New Orleans (January 8, 1815), fought in Louisiana's coastal wetlands, preserved the United States' control of the Mississippi Valley. (*New York Herald*, March 8, 1815)

the attendant jettisoning of its suspended heavy sediment load. Sustained deposits gradually form sedimentary lobes, and accretion broadcast over a broad geographic area resulting from the sporadic migration of the Mississippi's channels over centuries created a sprawling wetland stretching from the Atchafalaya River to the Rigolets. The Deltaic Plain's salient features include a highly irregular shoreline with associated rivers, natural levees, marshes, swamps, bayous, lakes, barrier islands, and salt domes.

Every successive Mississippi delta that contributed to the formation of the modern Deltaic Plain underwent several predictable stages of development. Over time, the gradient and carrying capacity of the lower river gradually decreased, and a shorter route to the sea eventually "captured" the main channel's flow. A diversion occurred, and the new course caused a shift in the location of the active delta's sedimentation regime. This shift left the old channel and delta to "die" and become yet another building block in Louisiana's burgeoning coastal wetland.

In active deltas, distributaries—watercourses drawing water away from the trunk channel—provided the mechanism for advancement of the principal lobes and sublobes into the shallow waters of the continental shelf. Distributaries inevitably branched and subdivided, thereby accelerating the distribution of river-borne sediments within the coastal lowlands. Most contemporary authorities identify five mature Holocene delta complexes (Maringouin/Salé/Cypremort, Teche, St. Bernard, Lafourche, and Balize), while a sixth (Atchafalaya–Wax Lake) is in an early stage of development. The new landmasses created at rivers' mouths frame, protect, and shelter associated deltaic estuaries and wetlands. These geographical features are the Gulf Coast's first line of defense against all storm events, from minor squalls to wave action linked to cold fronts, to hurricane storm surges, to ongoing marine processes such as saltwater intrusion, tidal currents, and sediment transport.

The Mississippi's peripatetic quest for the steepest available outlet gradient has produced distinctive near-sea-level topographic features that are changing at varying rates tied individually to localized sedimentation cycles. These cycles consist of alternating periods of retreat and advance resulting from relative rises and declines in sea level. These evolutionary cycles are predictable—older deltas deteriorate faster than younger ones. This is seen most clearly in the Mississippi Delta. While the river's bird-foot delta is in rapid retreat, one of the world's fastest-growing deltas is expanding at the mouth of the Atchafalaya River and Wax Lake Outlet, the Mississippi's most significant distributary.

This striking geomorphic dichotomy, however, is not entirely the result of natural processes. Rather, it is an artifact of long-standing engineering efforts to control the

Mississippi's destructive annual floods. Responding to public outcries for immediate relief, politicians expeditiously adopted enabling legislation, setting in motion engineering projects of unprecedented scale to alter the natural water distribution patterns—particularly, recurring springtime floods. Tinkering with the system—without accounting for the long-term consequences of these mandated alterations—has resulted in a radically modified, continuously shrinking, and increasingly vulnerable landscape.

To comprehend fully the consequences of interrupting natural physical processes, one must first understand the environmentally dictated settlement patterns in Louisiana's coastal zone. From prehistoric to modern times, Louisianians have consciously sought out the relative safety afforded by "high" ground—the natural levees paralleling the region's rivers, creeks, streams, and bayous. In South Louisiana, the land nearest a waterway may be fifteen to twenty feet above sea level. There is a steep gradient, however, behind the natural levees, which descend rapidly to the backlands and *brûlés* (swamp ridges traditionally cleared by means of slash-and-burn methods). These sloping lands—now commonly delimited by the tree line marking the nearest boundary of the swampy backlands—typically have a surveyed height of less than three feet.

Before human manipulation of the natural hydrological system, the Mississippi annually deposited coarse sediments along the river's main channels as well as the associated network of distributary waterways. This depositional sequence elevated the land through vertical accretion along the channel margins into natural ridges commonly called natural levees. The finer materials, largely silts and clays, remained suspended longer and were carried by the floodwaters off the natural levees into the "backlands" that evolved into swamps and marshes. Swamps and marshes, in turn, influence proximate sedimentation patterns and contribute to the localized accumulation of organic mud. The various landscapes created by the river's sediment-laden floodwaters are sandwiched between the Mississippi's historic complexes of distributary basins. Geological evidence of the delta system's development and ongoing change is clearly visible in aerial imagery and historic maps.

Scattered throughout this landscape are relatively high areas formed by the remnants of ancient natural levees. These modest ridges are often no more than two to three feet above sea level. The localized concept of "high" land is thus relative, but, in South Louisiana, any modest elevation serves as the definitive benchmark for settlement. Even on high ground, however, human settlements are vulnerable to the ever-present threat of flooding.

Since the beginnings of colonization, this looming hazard has profoundly influenced the evolution of the regional vernacular architecture. The oldest surviving structures in lower Louisiana are consistently elevated, either on piers or perched above first-story "basements." Ground-level rooms, often used for storage, were designed to be expendable, while the second-story living quarters sat safely above floodwaters. This building style was once pervasive on natural levees, barrier islands, *chênières*, and "hammocks," which were the anchor points for each settlement node on the coastal plain.

The emergence of raised-cottage architectural styles was but one manifestation of the pragmatism and adaptability of the *petits habitats* (small farmers) and others who settled along the lower Mississippi and its distributaries. During the eighteenth century, Louisiana's French and Spanish colonial governments mandated the construction of modest levees, often no more than three feet in height. These earthen dikes—maintained by property owners, not governmental entities—naturally provided only a modicum of protection, particularly in sparsely settled areas where gaps in the levee system exposed everyone to heightened flooding risks. In addition, channelization of the river resulting from levee construction meant that sedimentation, confined to the riverbed, effectively guaranteed rising river levels and the settlers' consequent need to continuously maintain levees and periodically raise their heights in their doomed perennial campaign to keep the region's waterways in check.

Settlers had little choice in the matter, because topography not only dictated settlement patterns but also profoundly shaped the regional economy. Arable lands were concentrated in the natural levees, and South Louisiana's navigable waterways were the area's transportation and communications lifelines. Thus wedded to the natural levees, resilient South Louisianians faced the ever-present threat of flooding—even in urban settings, where greater individual and governmental resources were readily available.

Louisianians confronted this dilemma in the earliest days of colonization. For instance, in 1717 acting governor Jean-Baptiste Le Moyne de Bienville selected New Orleans's location over the objection of his chief engineer, Le Blond de la Tour, who cautioned that the proposed site was subject to frequent inundations. The engineer's dire predictions were painfully prescient, for floodwaters actually delayed construction of the Crescent City until 1718. This delay should have signaled Louisiana's colonial administrators that this nascent community—located at sea level and, later, as a result of subsidence, below sea level—would have to cope with the persistent threat of flooding, perhaps best illustrated by the frequent historic crevasses—that is, levee breaks—inundating lands contiguous to the Mississippi, including metropolitan New Orleans.

From 1801 to 1927 forty-seven major flooding events—one nearly every three years—devastated Louisiana. Of these natural disasters, the 1882 and 1927 floods are most notable. Identified by contemporaries as the greatest peacetime disaster in the nation's history, the 1927 flood inundated parts of seven states collectively equal in size to the state of Illinois. The flooding emanated from two sources: natural sheet flow and crevasses. Thirteen crevasses appeared along the lower Mississippi during the 1927 flood, and each levee breach siphoned off enough surging water to reduce flood stages downstream.

The disaster constituted a watershed in US environmental history, for it provided the political impetus for implementation of the levee-only policy that still dominates the nation's flood-control strategy along the Mississippi River. This approach has had catastrophic long-term consequences for the very communities it was designed to protect.

Before 1928, when construction of the gargantuan engineered levees that currently line the major waterways of the lower Mississippi Valley began, crevasses were natural phenomena that effectively maintained the original regional hydrology and sedimentation patterns. Because local resources were overwhelmed by the challenge of restoring damaged levees, many breaches remained unrepaired for years. The resulting gaps in the levee system essentially functioned as high-water safety valves, whose sediment-heavy, torrential discharges replenished cultivable lands, thereby permitting farmers and plantation owners to maintain the practice of monocropping—a system of high-yield agriculture in which the same crop is planted year after year.

While sediment-laden floodwaters from sheet flow and crevasses maintained the viability of natural-levee agriculture, discharges associated with the Mississippi's historic meanderings in the delta also sustained the swamps and marshes that were entirely dependent upon freshwater and nutrients provided by the river system. The Mississippi's complex network of distributary channels and their sediment-deposition regimes produced the peat mats that sustained Louisiana's cypress and tupelo swamps. After filtering through the swamps, these nutrient-rich waters nurtured the coastal marsh—a landscape element extending approximately 300 miles along the coast and as much as 80 miles inland. These coastal marshes represent about 40 percent of the wetlands of the continental United States. Scholars have divided this landmass into nine geographical basins—Atchafalaya, Barataria, Breton Sound, Calcasieu-Sabine, Mermentau, Mississippi River, Pontchartrain, Terrebonne, and Teche-Vermilion (in some cases the Pearl River basin is added to the discussion). The health of these basins is tied directly to the current availability of Mississippi River

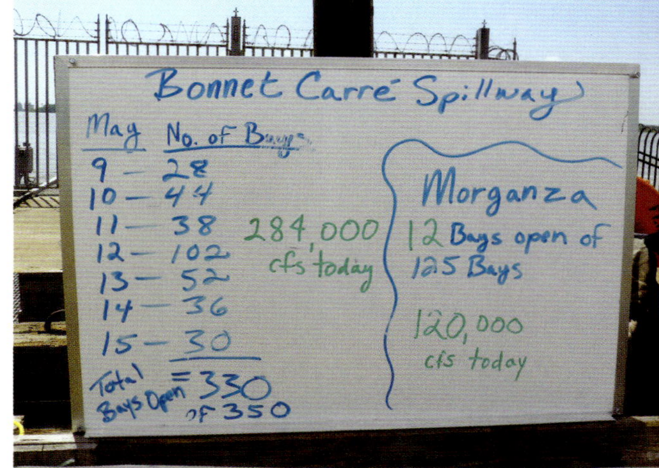

Located approximately thirty miles upriver from New Orleans, the Bonnet Carré Spillway (completed in 1931) was opened on May 9, 2011. By May 15, 330 of the Spillway's 350 bays were opened, with the water quickly spreading across Lake Pontchartrain. The 2011 event marked the tenth time the spillway's pins were removed from the structure's gates. The eleventh opening occurred in January 2016, when floodwaters were diverted for twenty-two days. (Photo by the authors, 2011)

By November 29, 1927, seven months after explosions ripped open the protection levee, large dredges were at work repairing the crevasse at Caernarvon. (Photo courtesy of the National Archives and Records Administration, Prints & Photographs Division, College Park, MD, photographer unknown, 1927, call number RG 77-MRC, Box 2)

sediments. As a result of the wholesale disruption of the region's natural hydrology, most parts of the coastal marsh are receding, some sections are stable, and a few others are growing, particularly in the Atchafalaya delta.

The coastal plain's rapidly declining state of health is cause for alarm. Huge, once-vibrant cypress stands—former bulwarks against hurricane tidal surges—are nothing more than botanical cemeteries in which the withered, graying trunks, killed by saltwater intrusion, stand like headstones. Just beyond the moribund cypress groves, formerly pristine areas of coastal marsh are now great expanses of open water. The rapid, ongoing destruction of marshland signals the loss of the region's single most important natural hurricane buffer for the densely populated natural levees.

The threat to population centers is further exacerbated by the rapid disintegration of critical barrier islands, which were also sustained by natural hydrology. The virtual disappearance of these essential buffers exposed the coastal marshes more directly to the onslaughts of tropical weather systems and their hugely destructive tidal surges.

Barrier islands, marshes, and swamps are proverbial mineshaft canaries for coastal Louisiana. These neighboring and complementary environments—one without trees, the other with a dense canopy of forest cover—are key features of South Louisiana's physical geography. And their viability may ultimately constitute the critical litmus test for humankind's continuing ability to occupy the coastal plain.

To the untrained eye, the marsh—in the rare instances where it retains its original structural integrity—is an uninterrupted, uniform sea of grass. In reality, however, these coastal grasslands are divided by salinity levels into four zones—saline, brackish, intermediate, and fresh. These ecological zones girdle the Louisiana coast in concentric bands, with the saline marsh bordering the Gulf of Mexico, the freshwater marsh adjoining the natural levees, and the Pleistocene "upland." The remaining salinity belts are sandwiched between these features. The freshwater marsh is further distinguished from its more salty cousins by the presence of high organic matter. This is particularly true of the Deltaic Plain, where high concentrations of water-borne organic matter have created the so-called trembling marsh—locally identified as *flotant* or *prairie tremblant*.

Before 1928, the salinity levels critical to the survival of the trembling marsh were naturally regulated by freshwater flow through swamps that, in the Louisiana coastal plain, consist primarily of dense groves of bald cypress and water tupelo. Freshwater swamps typically form one-to-two-mile-wide bands contiguous to streams. However, along the Atchafalaya River, the Mississippi's biggest distributary, the swamp encompasses nearly one million acres, making it North America's largest bottomland swamp.

The Atchafalaya Basin measures approximately 140 miles in length (roughly the distance between Washington, DC, and Philadelphia). This "river of trees" is one of the nation's last great wildernesses, and its southern terminus forms one of the world's most active and rapidly expanding deltas. Bounded on the east and west, respectively, by the natural levees of the Mississippi River and Bayou Teche, the basin's diverse biomes serve as filters for the huge volume of sediment funneled through the interior wetland by a massive and extensive levee system.

To the west of the Atchafalaya Basin, beyond Bayou Teche, stretches the final major feature of Louisiana's coastal plain—the vast interior prairies. Extending from the 31st longitudinal meridian to the Rio Grande, this region consists of naturally treeless expanses, punctuated by ribbons of woodland along rivers, bayous, and creeks or gullies, known locally as *coulées*. Underlying the open lands—originally grasslands—is a thick, dense clay formation that thwarted tree growth. Southwestern Louisiana's prairies once encompassed approximately 2.5 million acres (about two-and-a-half times the size of present-day Houston), but less than 1,000 acres of undisturbed prairie land (0.0004 percent) remain.

Cognizant of the constant danger of flooding, the New Orleans Levee Board, in December 1898, lined up pile drivers along the waterfront to reinforce the levee within the third municipal district—the area called Faubourg Marigny. (From the authors' collections)

Louisiana's prairie region is the critically important freshwater source for the *Chênière* Plain west of Marsh Island, but here, as elsewhere in the coastal region, the landscape is under intense pressure from environmental change and degradation. The prairie region's challenges, however, pale in comparison to those of the ecological zones in closer proximity to the Gulf. The coastal plain's marshlands, among the planet's youngest landforms, are eroding and subsiding at an alarming rate. Indeed, since 1930 Louisiana has lost a landmass comparable in size to the state of Delaware.

Increasingly exposed to inherently destructive physical forces and cut off from life-giving nutrients by the diversion of sediments to the edge of the continental shelf by the guide levee system, Louisiana's coastal region is manifestly ill-prepared to face new challenges in the form of projected twenty-first-century sea-level rises. However, to the people of this region, radical environmental change constitutes just another complication to their lives.

Coastal denizens are unquestionably the key to any effort to explore the history and geography of the coastal lowlands. Yet, for too long, persons and entities who control the region's destiny have deemed the human population of this landscape less important than either corporate interests—and their respective economic concerns—or even "critters"—whether endangered or not. In order to understand the basis for this situation, one must first comprehend the historical processes that have contributed to this state of affairs.

A NEW MAP of the RIVER MISSISSIPI from the SEA to BAYAGOULAS
LAKE PONTCHARTRAIN
Lake Maurepas
Manchac
Bayagoulas
Colapissas
Little Colapissas
Les Allemands or Carlstein
NEW ORLEANS
Bayouc St. John
LAKE OUACHAS
LAKE BORGNE
Low and Marshy Meadows
Shallow water with many Small Islands very little Known
Isles de la Chandeleur
Candlemas Islands
I. aux Vaisseaux Ship I.
I. aux Chats or Cats I.
I. aux Oiseaux Birds I.
La Bonne Chere
Chenaux
Bay St. Louis
Grand Bayouc
Ensenada de Palo or Wood's Bay According to the Spanish Charts
Ruins of Fort la Boulaye the First Settlement made in 1700
la Hache
I. au Breton
East Pass
La Balise
Entrance of the River
South Pass or Pass a Serigny
Cabo de Lodo or Mud Cape According to the Spanish Charts
South-West Pass
MOUTHS OF THE MISSISSIPI
GULF OF MEXICO
British Statute Miles
5 10 15

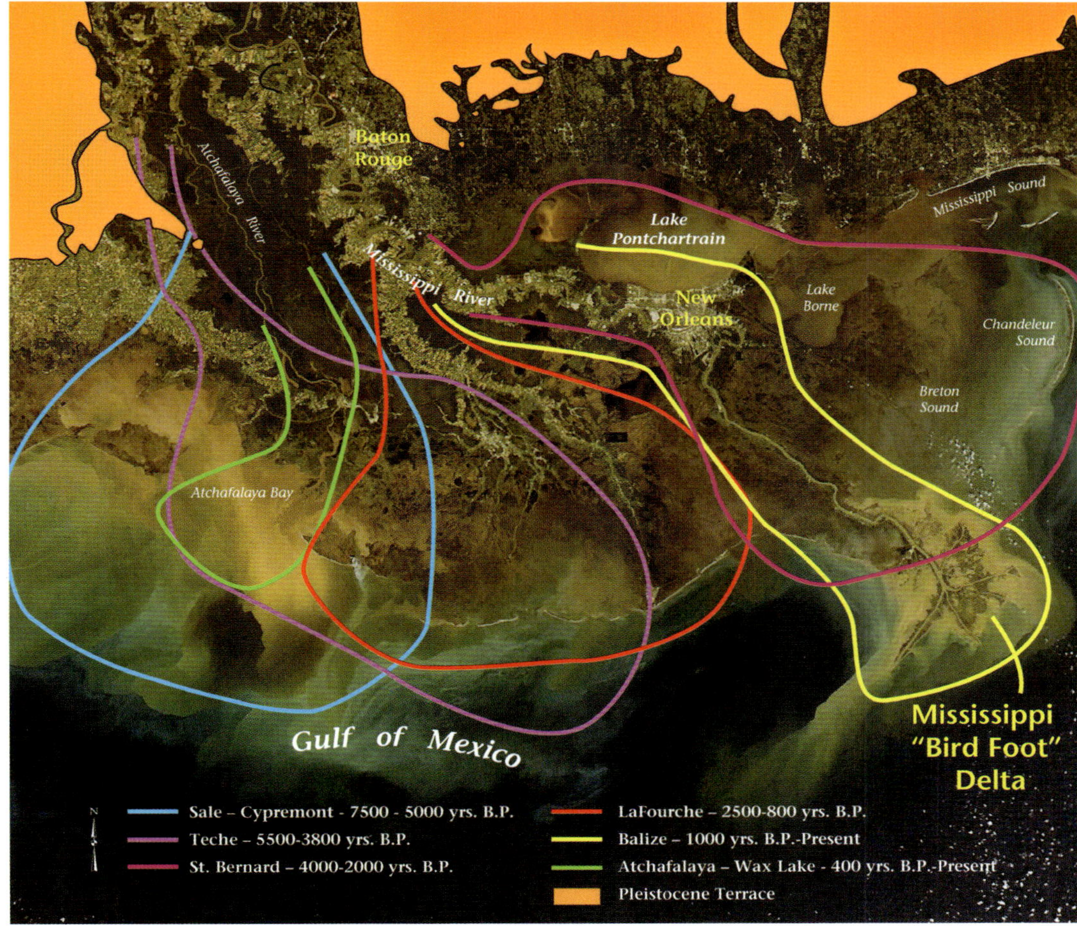

The Mississippi's Delta lobes modified from H. H. Roberts (1997) by Lisa Pond.

Nineteenth-century levees were modest earthworks, as this 1887 sketch clearly indicates. (Lithograph from W. W. Glazier, *Down the Great River: Embracing an Account of the Discovery of the True Source of the Mississippi* [Philadelphia: Hubbard Brothers])

Repetitive crevassing at this site led to the development of the Bonnet Carré Spillway. (Lithograph courtesy of the Center for Louisiana Studies, University of Louisiana at Lafayette, date unknown)

In the early twentieth century, Louisiana's coastal marshes consisted of a virtually unbroken carpet of native grasses stretching from the Sabine River to the Pearl River. The wholesale disruption of the region's "natural plumbing" in the wake of the disastrous 1927 flood, however, denied this fragile ecosystem the sediment and waterborne nutrients necessary for it to sustain itself. The result has been erosion on an unprecedented scale, as the grasslands wash away. (Photo by the authors, 1986)

"Here is the place where land is cheap and plenty. Untold miles of it are unoccupied by man, and many miles more are always in the making. As we were passing a sportsman's club house, the warden remarked that about twenty years ago this was on the coast; now it is four miles inland. This deposit of the river makes the richest kind of soil."

Source: H. K. Job, "Adrift in the Louisiana Marsh," *Outing* 57(4) (1910–1911): 454.

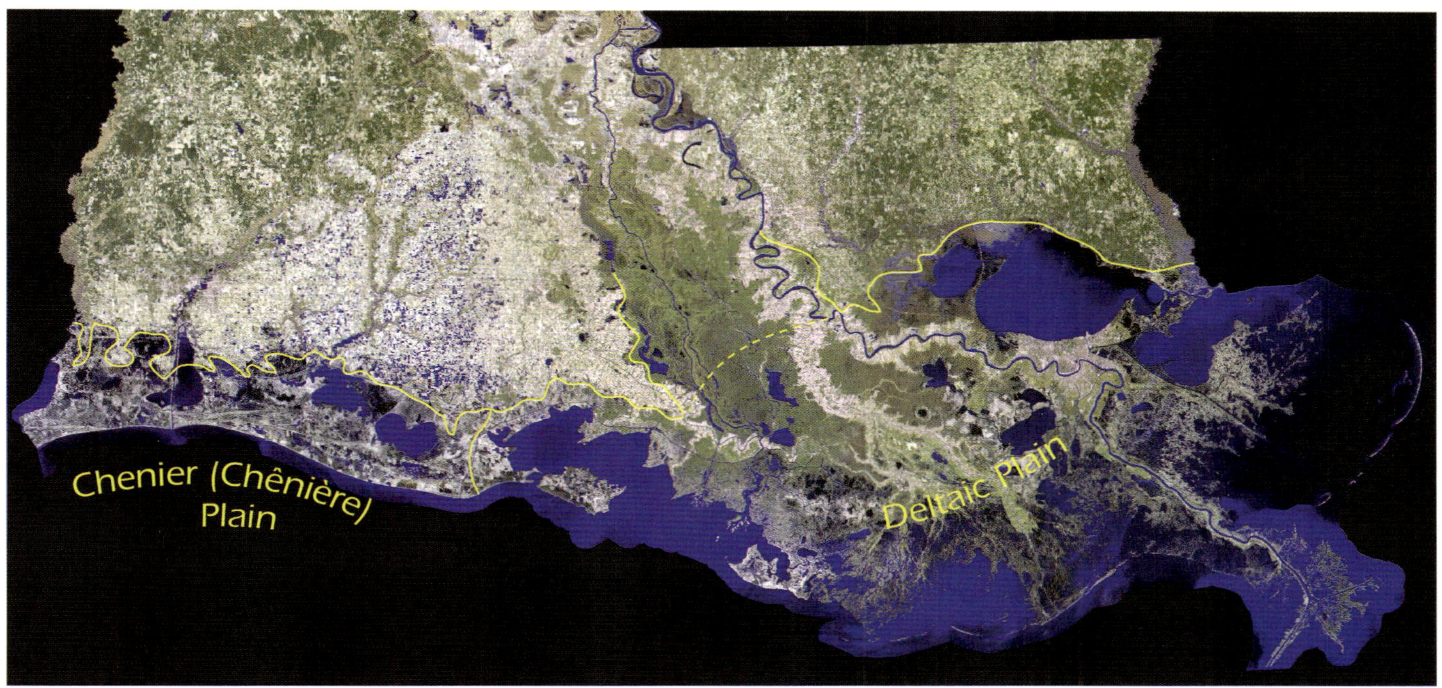

The *Chênière* and Deltaic Plains (Map compiled and edited by Lisa Pond, cartographer)

A Letter from Charles-Town, South-Carolina, dated March 18, ſays, " Advices from the Havannah inform us, that at length Don Antonio d'Ulloa had ſailed from thence with Troops under his Command, for the Miſſiſippi, and had actually taken Poſſeſſion of the City and Iſland of New Orleans and the Province of Louiſiana. The French Colony Troops remain, having entered into the Spaniſh Service." Theſe Letters add, that they have received an Account at Charles Town, that a Number of Families from Bermuda have lately arrived at Eaſt-Florida, in order to ſettle there, and more are expected.

On March 5, 1766, Antonio de Ulloa, Louisiana's first Spanish governor, took possession of New Orleans. Ulloa's successor, Alejandro O'Reilly, implemented land grant regulations that not only left a permanent imprint on coastal Louisiana settlement patterns but also instituted levee construction policies that resulted in the first tentative efforts to alter the region's natural "plumbing." (*London St. James Chronicle*, June 5, 1766)

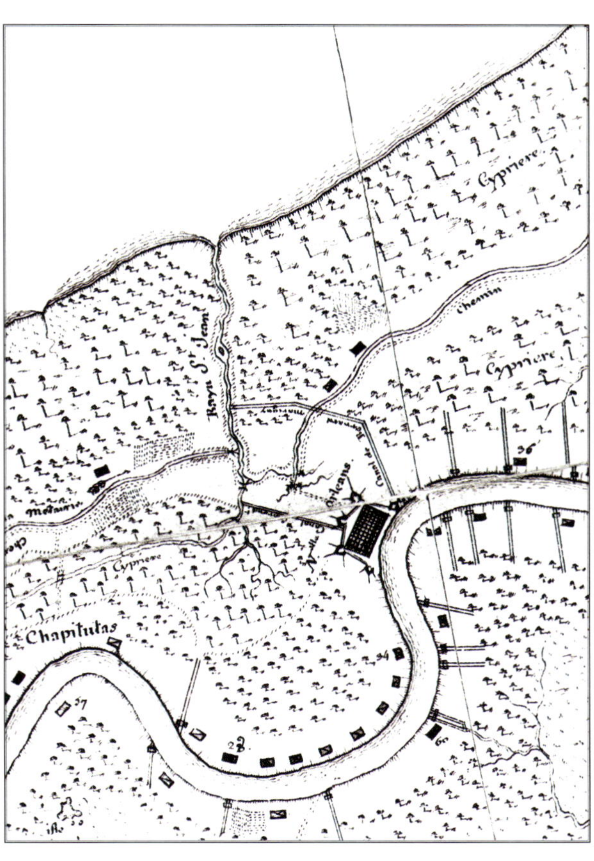

This map of the New Orleans area, drawn by engineer/surveyor Charles (Carlos) Trudeau, demonstrates graphically how early historical settlement was confined to the natural levees. (Map courtesy of the Center for Louisiana Studies, University of Louisiana at Lafayette, date unknown)

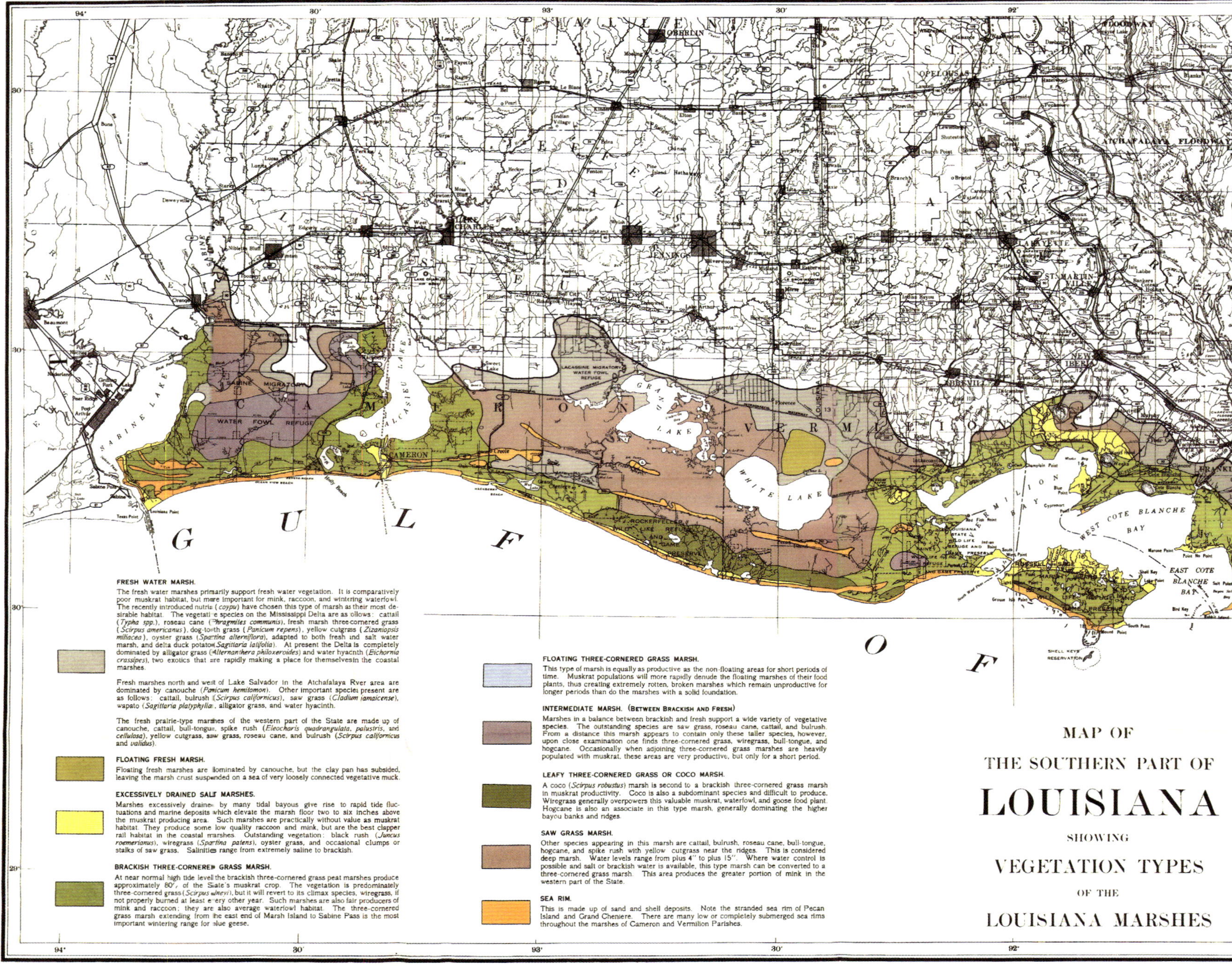

Along the Louisiana coast, the distribution of native flora coincided with salinity levels, and manmade alterations to the natura movement of water and sediments from major streams to the

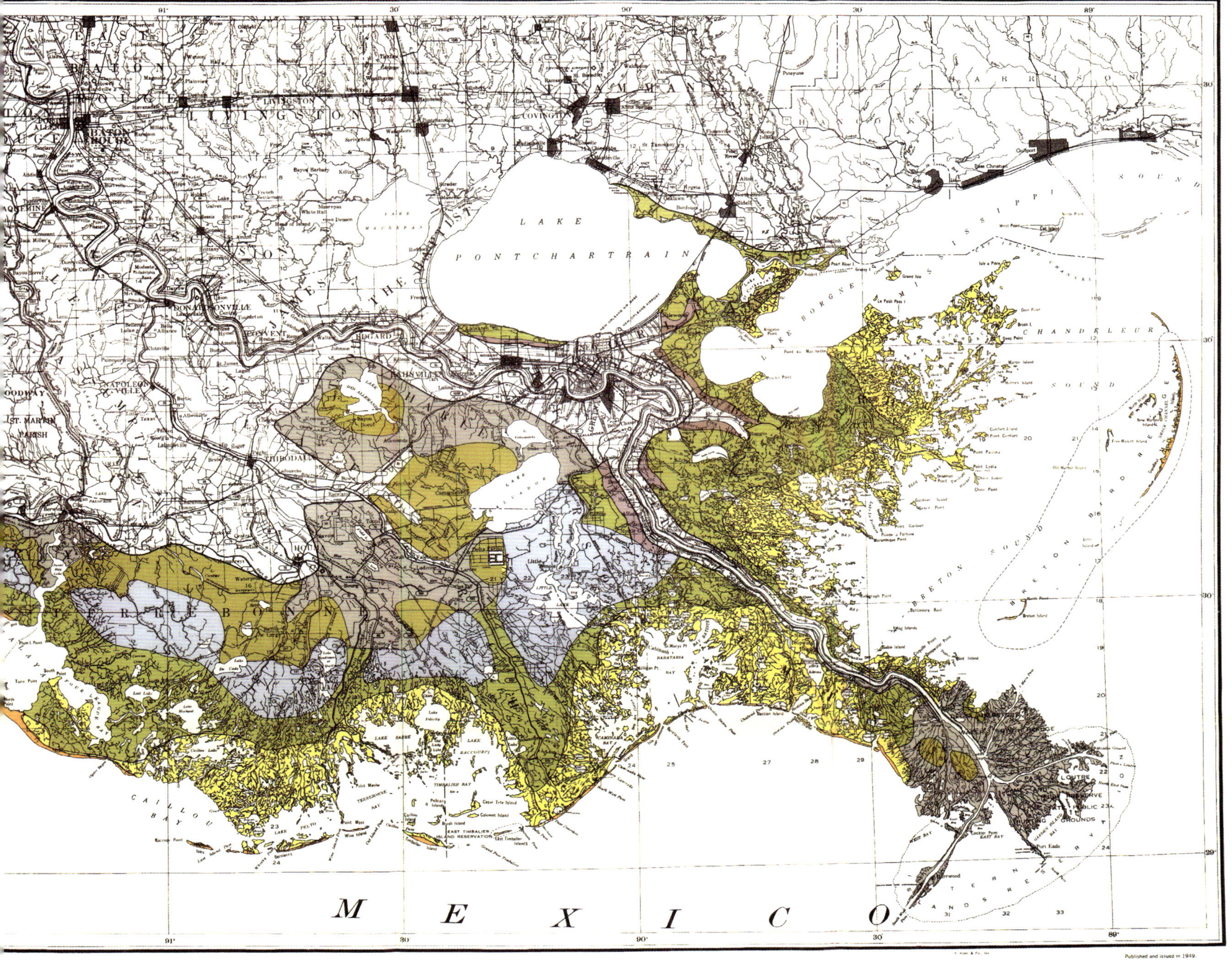

LAKE PONTCHARTRAIN
LAKE MAUREPAS
LAKE BORGNE
MISSISSIPPI SOUND
CHANDELEUR SOUND
BRETON SOUND
MEXICO
LIVINGSTON
ST. TAMMANY
HARRISON
HANCOCK
TERREBONNE
BARATARIA BAY
CAILLOU BAY
LAKE PELTO
TERREBONNE BAY
TIMBALIER BAY
LAKE SALVADOR
LAKE CATAOUATCHE
LAKE RACCOURCI
LAKE BARRE
BATON ROUGE
DONALDSONVILLE
THIBODAUX
HOUMA
NEW ORLEANS
COVINGTON
Gulfport
Biloxi
Pass Christian
CHANDELEUR ISLANDS
BRETON ISLAND
PASS A LOUTRE
EAST TIMBALIER ISLAND RESERVATION
91°
30'
90°
89°
Published and issued in 1949.

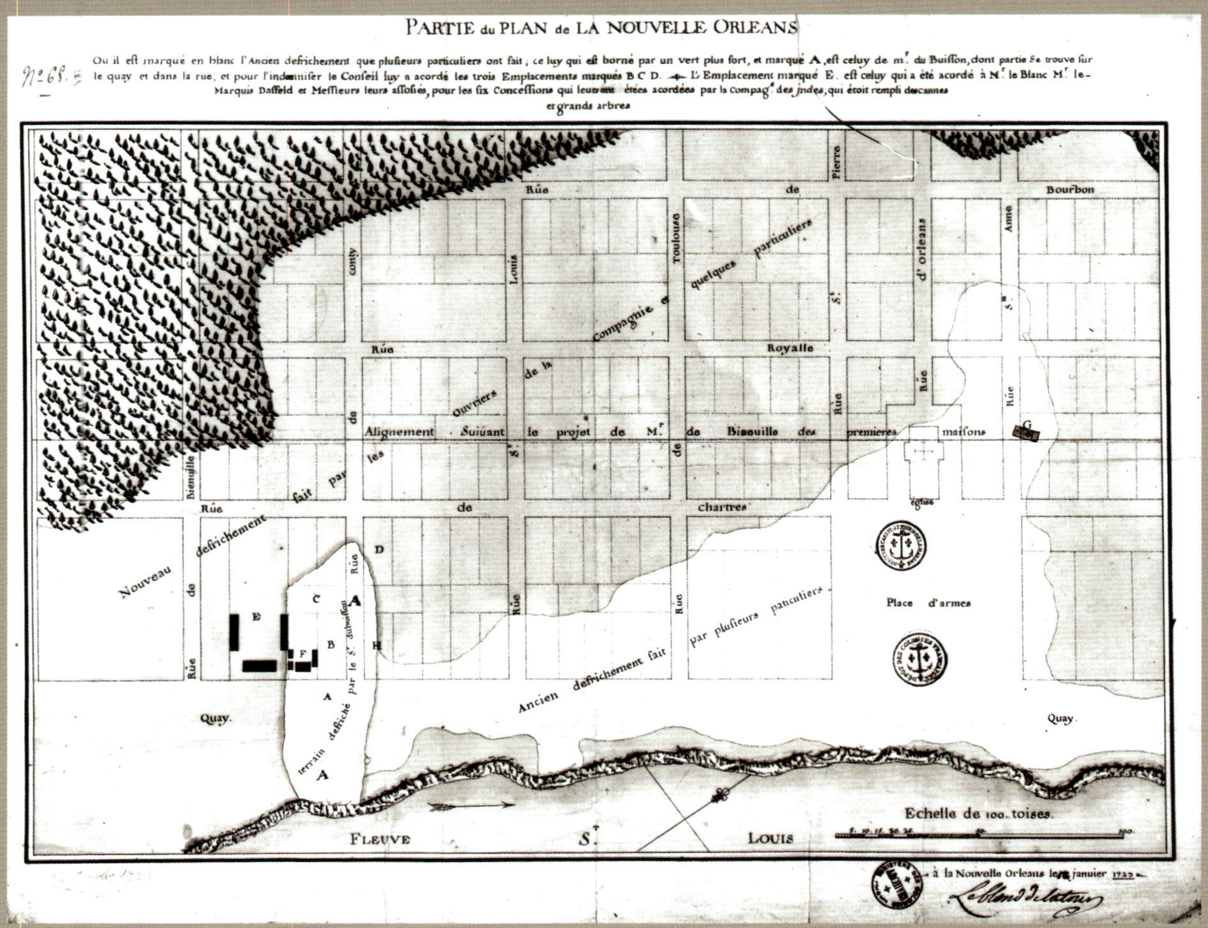

In 1723, when this map was created, New Orleans consisted of a tiny clearing encompassing approximately four city blocks on the Mississippi River's natural levee. The shaded area delineates the blocks yet to be cleared in the next phase of the Crescent City's development. (Map courtesy of the Center for Louisiana Studies, University of Louisiana at Lafayette)

In areas, like New Orleans, facing the threat of inundation from multiple fronts, communities usually had enough resources to deal with only the most immediate threat—the river. But as the quotation below makes abundantly clear, any defensive system is only as strong as its weakest links, and gaps or substandard levees upstream threatened the security of all communities downstream: "The city of New Orleans and its suburbs, with a population of 216,090 cover nearly all of the higher land lying within Orleans parish. The rear of [the] city itself almost touches the swamp land, originally timbered with cypress, passing into the marsh prairie that borders Lake Pontchartrain with a depth of from 3 to 4 miles. The Great Levee protects the city front from the flood-waters of the Mississippi; but the enemy not uncommonly finds its way to the rear, through breaks in levees above or below, not so well cared for."

Source: W. H. Harris, *Louisiana Products, Resources, and Attractions: With a Sketch of the Parishes: A Hand Book of Reliable Information Concerning the State* (New Orleans: New Orleans Democrat, 1881), 181.

Because a treacherous bar across the main navigational channel impeded access by oceangoing warships, Louisiana's colonial leaders considered the modest, poorly maintained fortifications at Balize, an island at the mouth of the Mississippi River, to be an effective first line of defense against enemy incursions. (Map courtesy of the National Archives and Records Administration, Prints & Photographs Division, College Park, MD, 1813, call number G4014. B16:2F61813.L32)

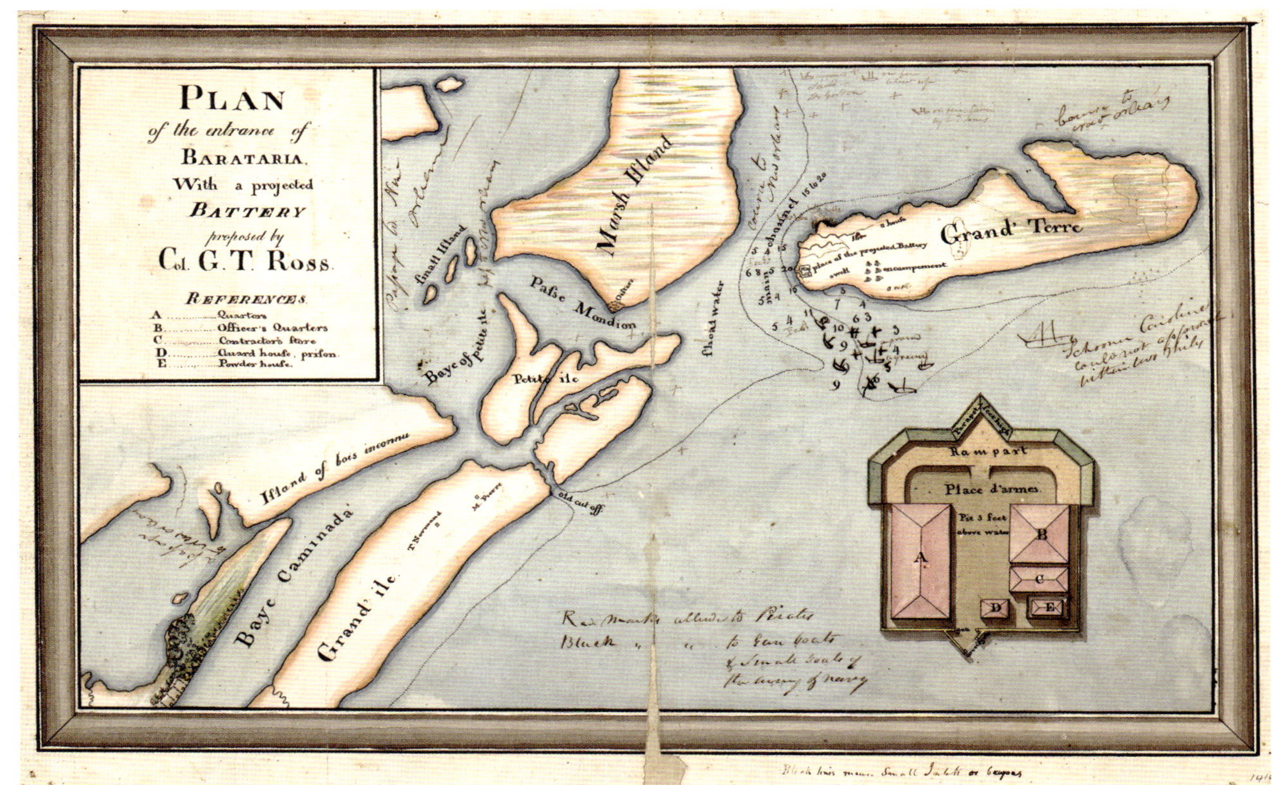

Louisiana's barrier islands have historically been coastal communities' first line of defense against hurricane tidal surges. During Louisiana's late territorial and early statehood periods (ca. 1809–1815), Grand Terre, an island at the mouth of Barataria Bay, served as the headquarters and operational base for Jean Lafitte's pirates. Once these buccaneers were eradicated, the United States' military built a major fortification—Fort Livingston—to guard the southern approaches to New Orleans. (Map courtesy of the National Archives and Records Administration, Prints & Photographs Division, College Park, MD, 1815, call number G4014.N5S42 1815.P4)

Louisiana's *Chênière* Plain. (Map compiled and edited by Lisa Pond, cartographer)

> "In a practical sense, the earliest land survey and engineering records of the French settlements in Louisiana are records of a great land reclamation project. Early manuscripts contain many details of floods and the need for levees; of high precipitation and the need for drains; of tidal action and the need for a sea wall; and of low elevation and the need for artificial (or pump) drainage."

Source: R. W. Harrison, *Alluvial Empire: A Study of State and Local Efforts toward Land Development in the Alluvial Valley of the Lower Mississippi River, Including Flood Control, Land Drainage, Land Clearing, Land Forming,* Vol. 1 (Delta Fund, 1961), 258.

The *Chênière* Plain does not stop at the Texas border. The Texas portion of this geologic feature includes the Anahuac, McFaddin, Texas Point, and Moody National Wildlife Refuges. (Photo by the authors, 2015)

Louisiana's shoreline is eroding at a rate of five to thirty feet per year. As a consequence, parts of Highway 82 have been moved repeatedly. The telephone pole in this image indicates that much of a former beach face had recently been replaced by open water. (Photo by the authors, 2009)

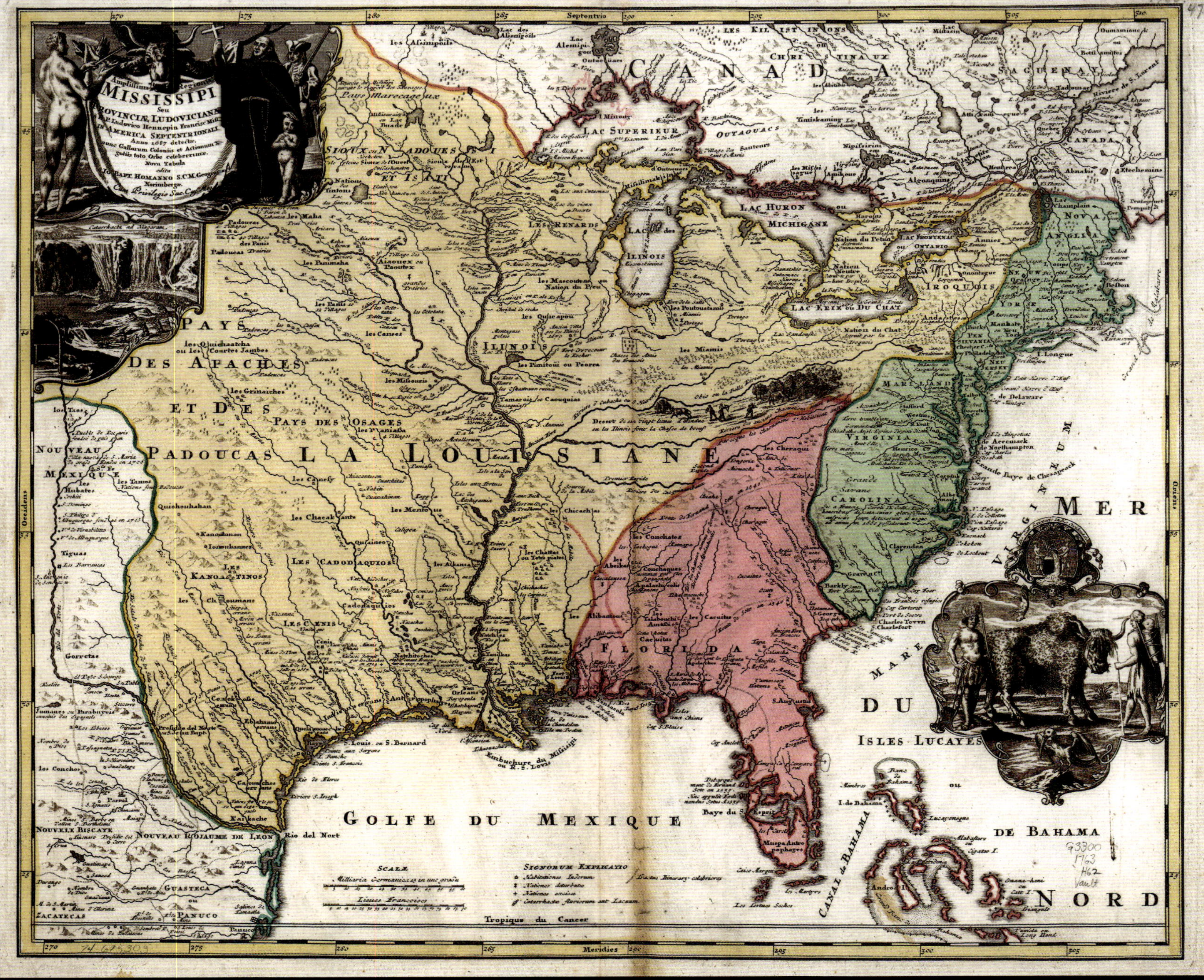

French Louisiana encompassed all of the lands drained by the Mississippi River. Because of the colony's vast expanse and primitive infrastructure, communications were possible only by means of waterways—waterways that all discharge in Louisiana's coastal wetlands. (Map courtesy of the National Archives and Records Administration, Prints & Photographs Division, College Park, MD, 1687, call number G4010 1759.H6)

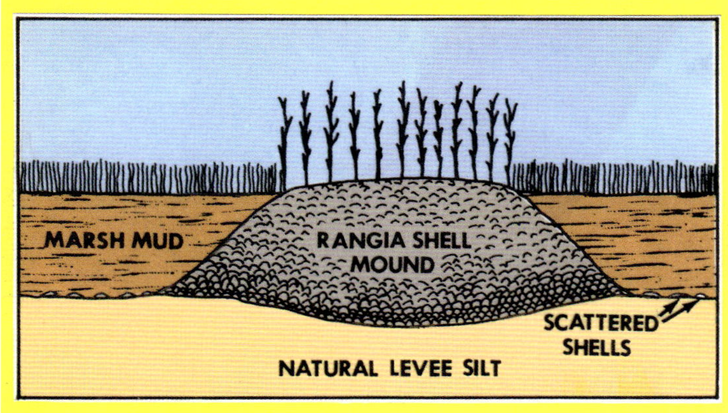

Middens are common archaeological features along the coast. These sites have served as fundamental landmarks for researchers unraveling the Mississippi River's deltaic history. (From the authors' collections, used with permission of Harley Jesse Walker, Boyd Professor, Louisiana State University)

THE LITTLE TEMPLE, ON THE RIGOLETS, NEAR NEW ORLEANS, LA.—FROM A SKETCH BY JAS. E. TAYLOR.

The Little Temple was perhaps the most imposing Native American structure in the Louisiana coastal plain. In the early nineteenth century, this massive *Rangia cuneata* shell mound, or "midden," located at the junction of Bayous Rigolets and Perot, was both a major landmark and an important auction site for Baratarian booty. Little Temple mound no longer exists, having been excavated for its component oyster shells after the Civil War. According to George Washington Cable, excavators had removed more than 300,000 barrels of shells by 1884. (From the authors' collections)

Over the past two centuries, Louisianians typically named major crevasses to memorialize the massive destruction they caused. The most notable historical crevasses include the 1849 Sauvé, the 1850 Bonnet Carré, and the 1884 Davis breaks, which were deemed by the contemporary press—and, indeed, the public at large—as major environmental disasters. More than a century later, these crevasse sites are still easily discernible from the air.

38 HARPER'S WEEKLY. VOLUME XXVII., NO. 1407.

THE SOUTH PASS JETTIES OF THE MISSISSIPPI.—DRAWN BY J. O. DAVIDSON.—[SEE PAGE 790.]

1. Port Eads Light-house. 2. Bird's-eye view of the Jetties. 3. Dredging steamer *G. W. R. Bayley*. 4. Dredging apparatus of *G. W. R. Bayley*. 5. Showing former and present Channel. 6. Seaward end of Jetties. 7. Mattress-making. 8. Section of Jetty 11,800 feet at sea. 9. Section of Jetty. 10. Channel made by Current—2, former depth. 11. Concrete Works. 12. Inside Willow Sand-barriers. 13. View within the Jetties.

In the wake of the Civil War, siltation rendered the shallow mouth of the Mississippi River increasingly unnavigable to many of the heavier iron-hulled ships entering the passes after 1865. James Buchanan Eads, a largely self-taught engineer who designed the first bridge across the Mississippi River, solved this problem by building jetties to scour channels of acceptable depths at the waterway's mouth. Eads financed the project with his own funds, with the proviso that he be reimbursed by the federal government when the Mississippi reached the targeted depth. This lithograph shows maintenance work on the Southwest Pass's jetties. (Lithograph from *Harper's Weekly*, December 8, 1883, 788)

The US Quartermaster Depot's transit shed and wharfs under construction at New Orleans in early November 1941. The historical presence of numerous government installations—both military (Forts Livingston, Jackson, St. Philip, and Pike, as well as naval air bases at New Orleans, Houma, and Lake Charles) and civilian (numerous coastal lighthouses, customs offices, engineering offices, and so forth)—along the Louisiana coastline is a testament to the region's national strategic and economic importance. (Photo courtesy of the National Archives and Records Administration, College Park, MD, photographer and date unknown, call number RG 77-RH, Box 92, image 122a)

Congress adopted the Flood Control Act of 1928 in response to the horribly destructive flood of the preceding year. This landmark legislation empowered the US Army Corps of Engineers to design and build flood-control projects along the nation's waterways. Manual laborers, like those seen here near Donaldsonville, constituted the backbone of the early building efforts. (Photo courtesy of the National Archives and Records Administration, Prints & Photographs Division, College Park, MD, photographer unknown, 1927, call number RG 77-MRC, Box 2, 1a)

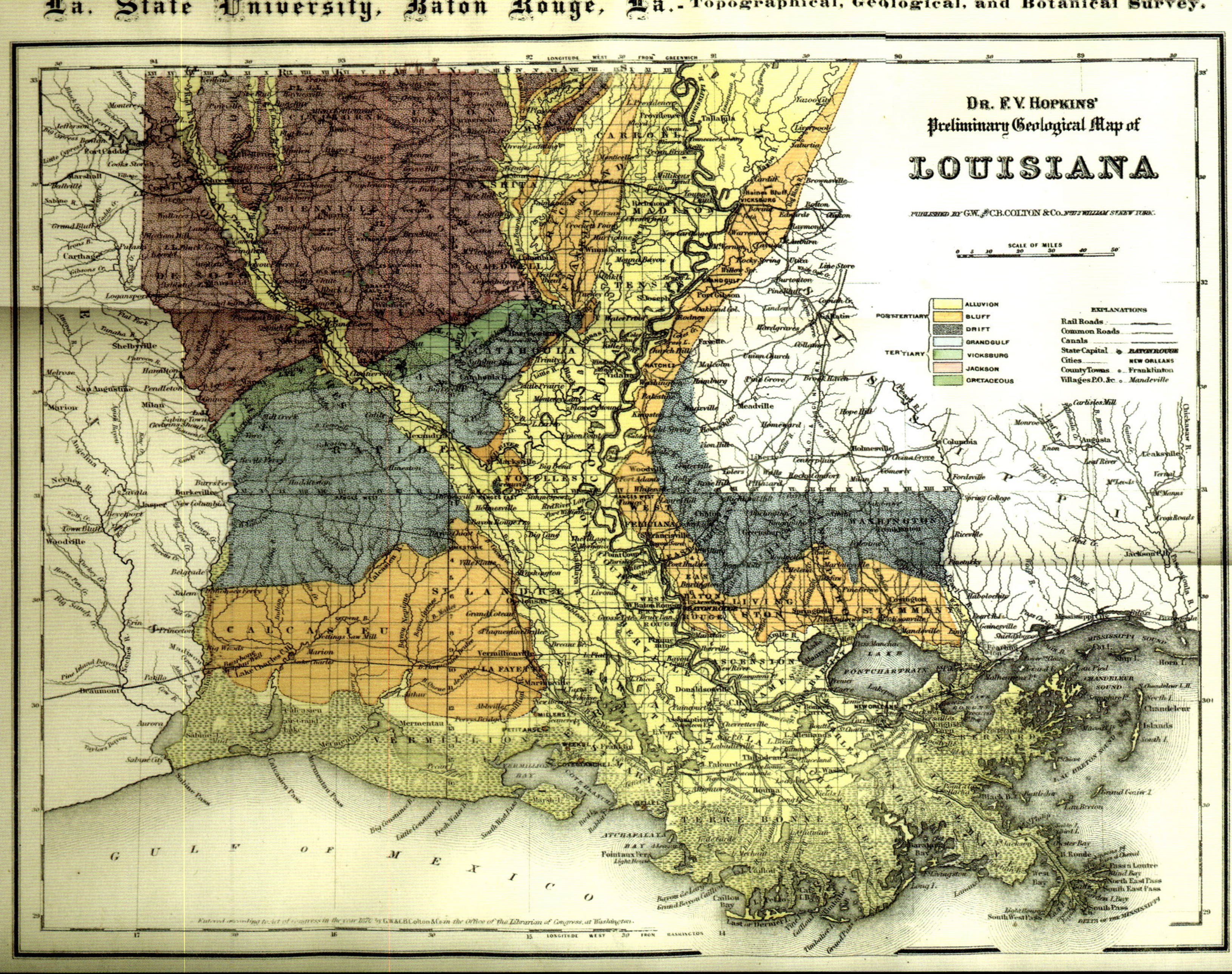

Published in 1870, F. V. Hopkins's map is the first to show Louisiana's geology. The Mississippi floodplain, the coastal marshes, and the Pleistocene uplands are clearly visible. (From the authors' collections)

Twenty-five years before the Great Hurricane of 1893, Chênière Caminada was a bustling fishing community. Only four homes survived the storm, and observers estimated the death toll at 779 to 822. (Lithograph from *Frank Leslie's Illustrated Newspaper*, August 1, 1868, 309)

August 1, 1868.] FRANK LESLIE'S ILLUSTRATED NEWSPAPER 309

SCENE IN THE ISLAND OF CHENIER CAMINADA, IN THE GULF OF MEXICO—MR. L. COLLINS, THE MAYOR, SCHOOLMASTER, AND "OLDEST INHABITANT," AND A GROUP OF HIS PUPILS AND PROTEGES.—FROM A SKETCH BY JAS. E. TAYLOR.

In 1898 Congress allotted funds for a survey of Southwest Pass in order to create a thirty-five-foot deep channel. Between 1899 and 1902, additional congressional allocations funded projects to improve the existing navigational channel. In the early twentieth century, field crews, working out of the Mississippi Delta community of Burrwood, were tasked with making channel improvements. By 1938 the community was a thriving settlement of over 1,000 residents. The now partially submerged village site is rapidly receding into the mists of obscurity. (Photo courtesy of the National Archives and Records Administration, Prints & Photographs Division, College Park, MD, photographer Russell Lee, 1938, call number LC-USF33-011810-M1)

The quotes that follow clearly indicate that a levee-only solution policy had crystallized in governmental federal flood control agencies decades before the 1927 flood: "All will welcome the day when we can return to this time-honored policy [of enriching the nation by gaining bloodless victories over natural forces], and when our military engineers shall again descend the Mississippi with no hostile intent, but charged with the duty of completing the work of protection, but building that magnificent system of levees which the Delta Report projects."

A sailboat in a canal extending from White Lake to Pecan Island. (Photo reprinted from *History of Vermilion Parish, Louisiana* [1983], courtesy Vermilion Historical Society)

"The official report of the Committee of Engineers to the Convention at Vicksburg in 1890, which represented the organized flood-control element of the whole valley, stated that the 'testimony of all engineers familiar with the subject' was that there was 'no engineering difficulty' in controlling the river by levees and that the levee system was the 'only agency' by which control could be accomplished. In 1912 the Louisiana Engineering Society, three hundred strong, and the Louisiana State Board of Engineers almost to a man urged levees only."

Sources: A. A. Humphreys and H. L. Abbot, *Report upon the Physics and Hydraulics of the Mississippi River: upon the Protection of the Alluvial Region against Overflow; and upon the Deepening of the Mouths . . . Submitted to the Bureau of Topographical Engineers, War Department, 1861* (No. 4) (Philadelphia: J. B. Lippincott, 1861), 41–42; Edwin Hale Abbott, *A Review of the Report upon the Physics and Hydraulics of the Mississippi River, Prepared by Capt. A. A. Humphreys and Lieut. H. L. Abbot, Corps of Topographical Engineers, United States Army* (Boston: Crosby and Nichols, 1862). A. D. W. Frank, *Development of the Federal Program of Flood Control on the Mississippi River* (New York: Columbia University Press, 1930), 122.

This 1863 map of the Atchafalaya Basin was prepared by Henry L. Abbot and is the earliest detailed map of the country's largest freshwater swamp. (Note the vast area encompassed by Grand Lake, which is now a tiny fraction of its original size.) A. A. Humphreys and Abbot spent months along the Mississippi River assessing the viability of the existing levee system. For nearly 100 years, their twin reports served as a guide to a levees-only policy in dealing with Mississippi River improvements. (Map from http://www.photolib.noaa.gov/htmls/cgs05237.htm)

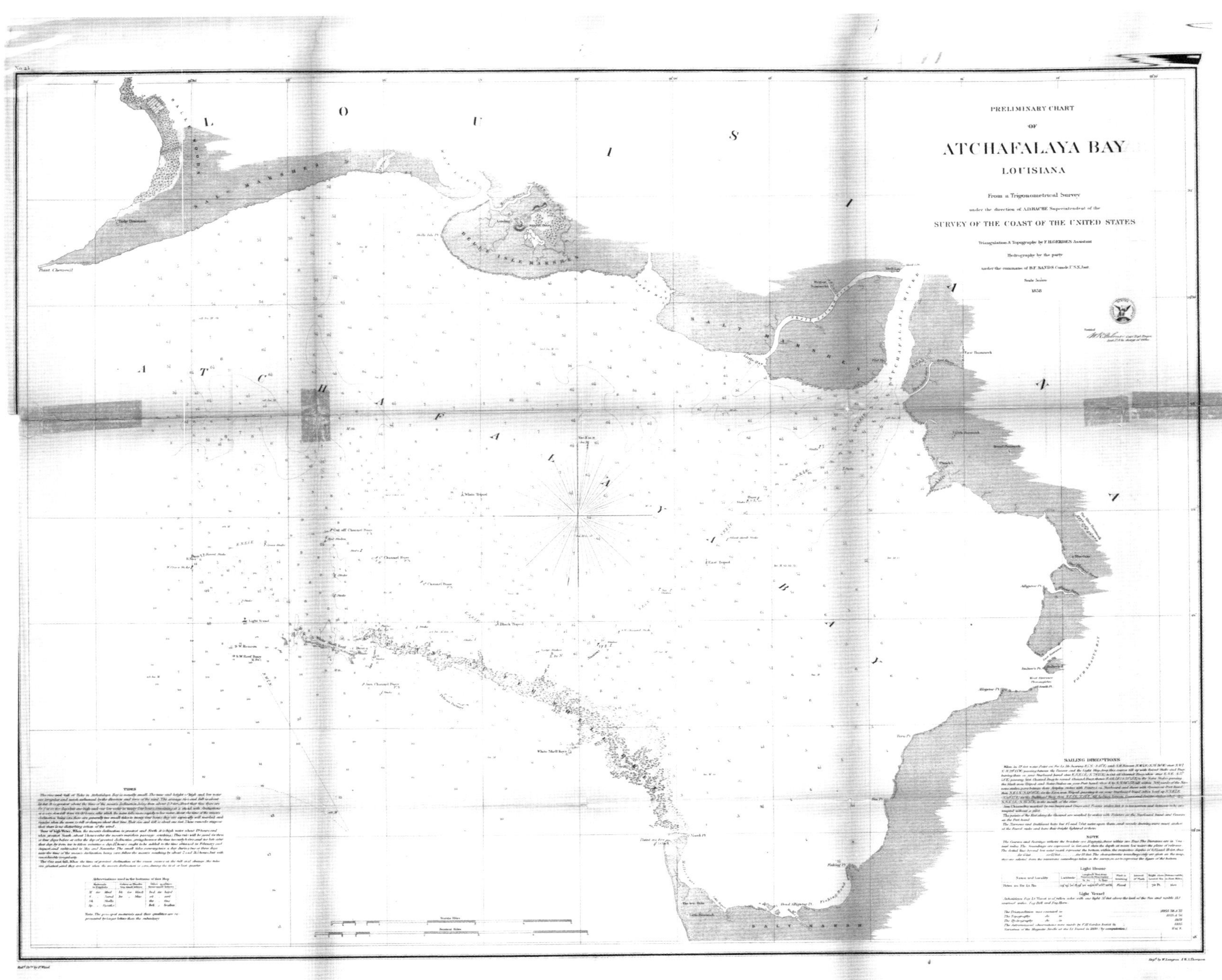

Atchafalaya Bay (now identified as West and East Cote Blanche Bays) constitutes the physical boundary between the *Chênière* and Delta Plains. This 1858 survey shows the bay's southern boundary, marked by this extensive reef system, trending west from Point Au Fer toward Marsh Island. Since 1914, dredges, whose efficiency has improved greatly over time, have harvested this massive natural shell reef for use in construction projects. Over time, dredging has generated tremendous public debate. (From the authors' collections)

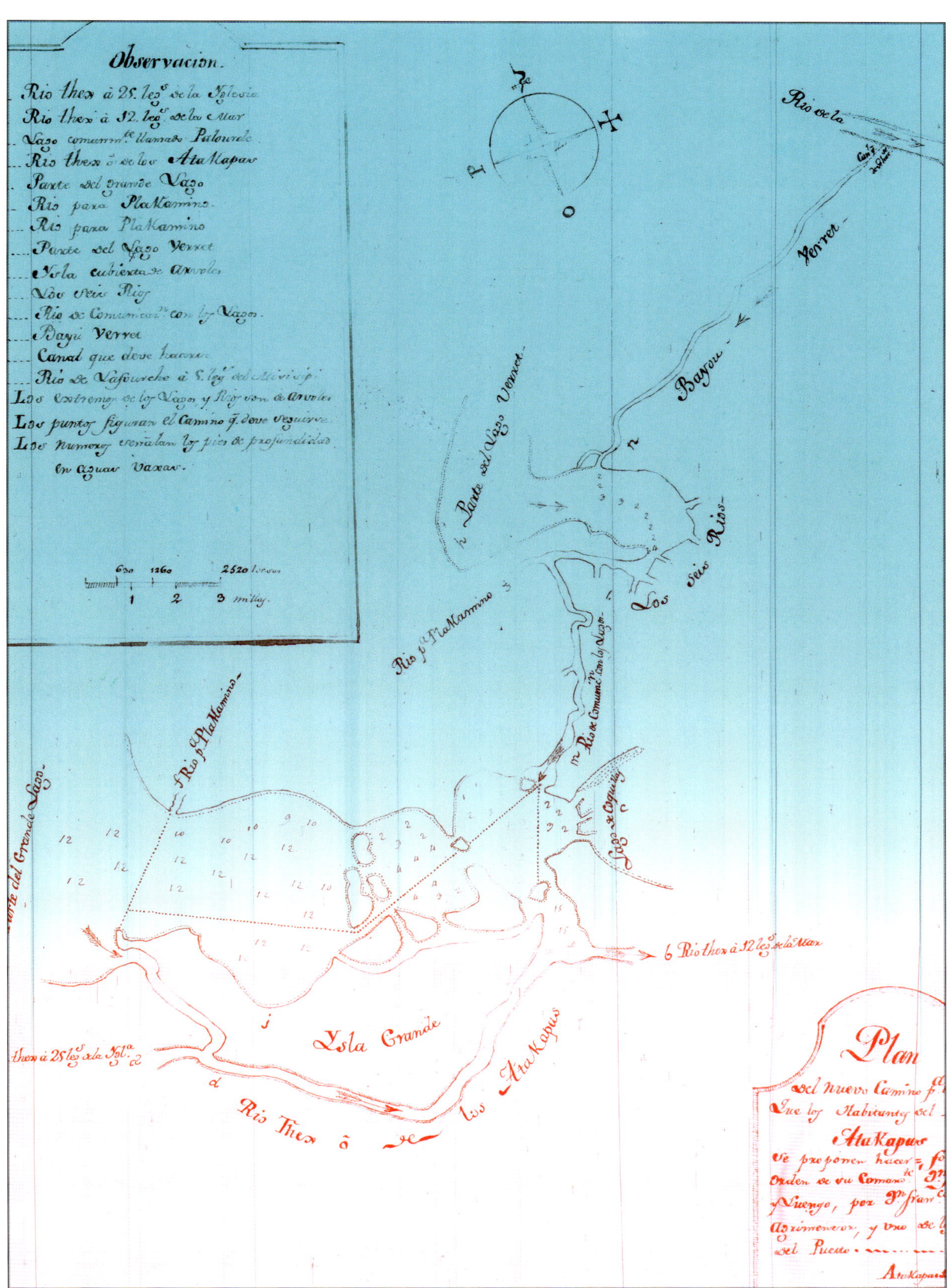

Eighteenth-century map showing the original hydrology of the lower Atchafalaya Basin. (From the Archivo General de Indias, Seville, Spain, P 29-43, Edicion 89-351)

"The reasoning of the 'levees only' people was simple and direct. They asserted openly that confinement presented not only the best way to control floods on the Mississippi, but the *only* practicable way. They maintained that all other plans had been tried in Europe and had failed. No workable plan of flood control could be devised, in their opinion, but to build levees 'sufficiently high' and strong enough to hold. They pointed with pride to the decreasing number of crevasses and the decreasing amount of lands inundated by each great flood. The crevasses occurred not because of any fault of the plan but because the plan had not been completed."

Source: A. D. W. Frank, *Development of the Federal Program of Flood Control on the Mississippi River* (New York: Columbia University Press, 1930), 123–124.

Less than a decade after the Civil War, when most of the Mississippi levees were still in ruin, the streets of New Orleans were often best suited for gondoliers using push poles. (Lithograph from *Every Saturday*, June 8, 1871, 41)

Indigenous architecture (as depicted in the elevations of these Creole-style buildings) emerged in the colonial era as a direct response to local environmental conditions. (From the Archivo General de Indias, Seville, Spain, P 29-43, Edicion 89-119)

Backbreaking labor was required to construct the 3,000-foot-wide channel associated with the 1911 Torres Crevasse. The break probably resulted from crawfish holes that weakened the levee's integrity. At the time, there was some concern that water from the crevasse would flood the low land from the east bank of the Atchafalaya to the west bank of Bayou Lafourche. (Photo courtesy of the National Archives and Records Administration, Prints & Photographs Division, College Park, MD, photographer and date unknown, call number RG 77-MRC, Box 2, Torres crevasse, image 88605-967a)

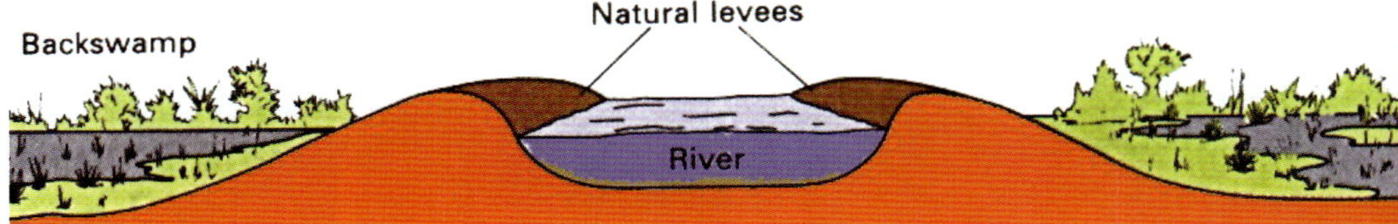

The natural levee frames a river's channel. These sedimentary features, which are aggregates of alluvial sediments, provide earth scientists with insights into flooding patterns. (Image from https://myweb.rollins.edu/jsiry/Seashore.html)

In the early 1930s operating engineers were at work building and rebuilding the Mississippi River's levees. (From the authors' collections)

"Coastal Louisiana has undergone a net change in land area of about 1,883 square miles from 1932 to 2010. This net change in land area amounts to a decrease of about 25 percent of the 1932 land area. Persistent losses account for 95 percent of this land area decrease; the remainder are areas that have converted to water but have not yet exhibited the persistence necessary to be classified as 'loss.' Trend analyses from 1985 to 2010 show a wetland loss rate of 16.57 mi^2 per year. If this loss were to occur at a constant rate, it would equate to Louisiana losing an area the size of one football field per hour."

Source: B. R. Couvillion, J. A. Barras, G. D. Steyer, W. Sleavin, M. Fischer, H. Beck, and D. Heckman, *Land Area Change in Coastal Louisiana (1932 to 2010)* (Washington, DC: US Department of the Interior, US Geological Survey, 2011), 1.

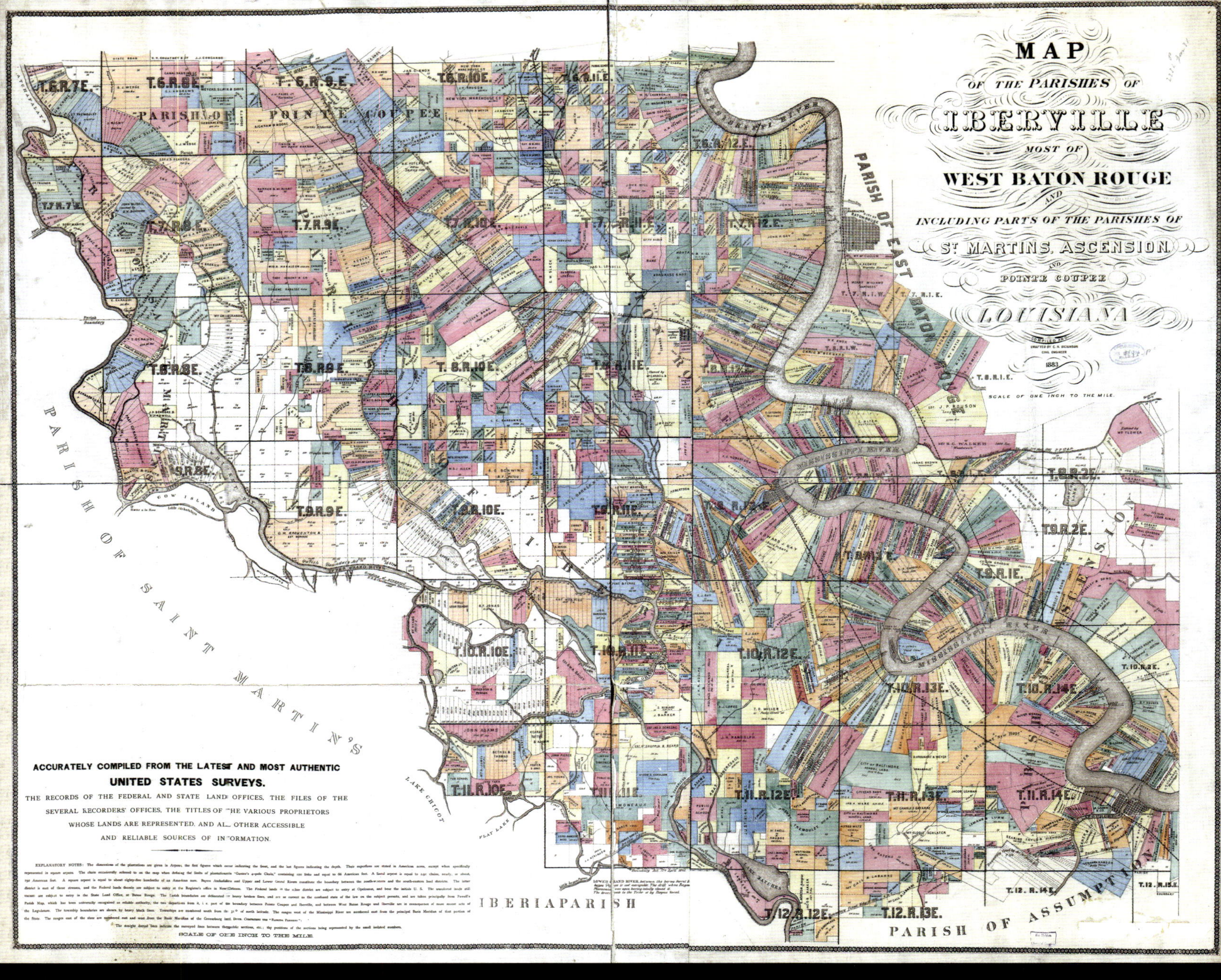

In Louisiana, unlike many areas of the country, there are three land division systems, all associated with the state's colonial history: arpent, a French unit of measure with landholdings fronting on waterways; metes and bounds, an English-derived method of describing land by physical features, such as boundary stones or witness trees; and township and range surveys based on rectangular grids. (Map courtesy of the Library of Congress, call number G4013.I3 1883 .D5, https://www.loc.gov/item/2011588002)

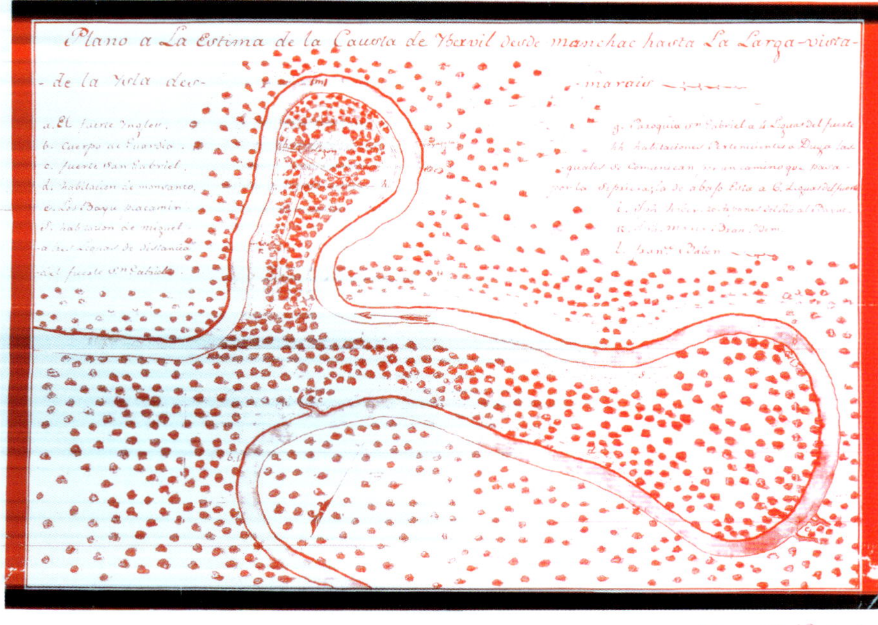

This Spanish colonial map indicates that all of the initial farmsteads in the Iberville District's St. Gabriel settlement lined the banks of the Mississippi River. (From the Archivo General de Indias, Seville, Spain, P 29-43, Edicion 89-162)

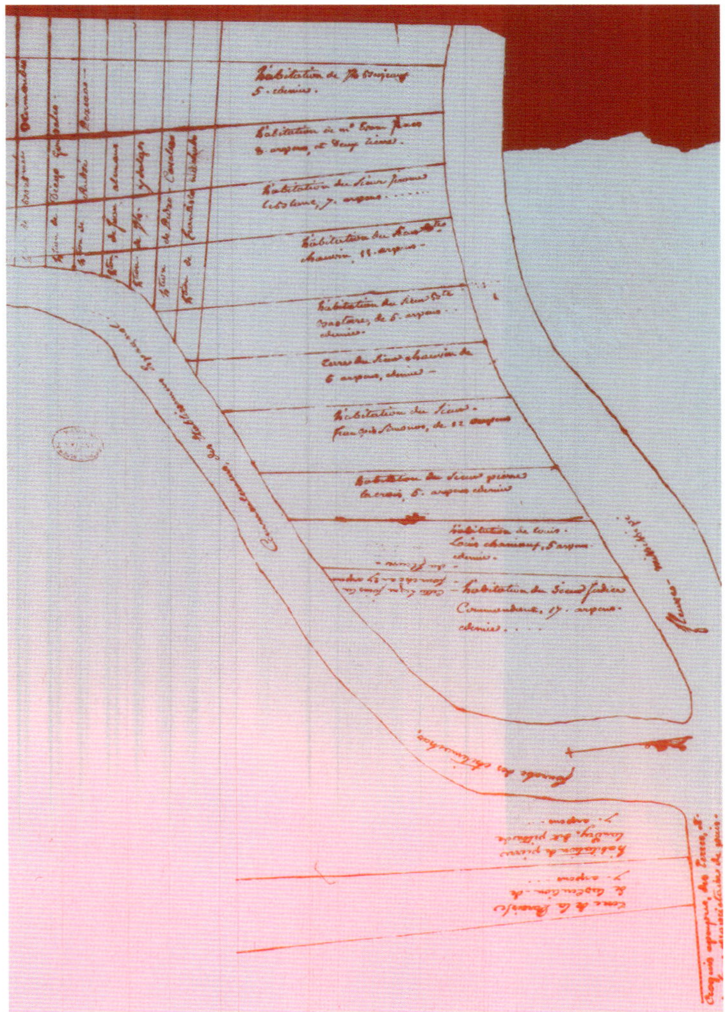

Spanish land grants at the junction of Bayou Lafourche and the Mississippi River. (From the Archivo General de Indias, Seville, Spain, P 29-43, Edicion 89-179)

This image shows clearly the effort required to protect riverfront communities from the 1927 flood. (Photo courtesy of the National Archives and Records Administration, Prints & Photographs Division, College Park, MD, photographer unknown, 1927, call number RG 77-MRC, Box 1, Folder 3-1)

Repairing a Mississippi River crevasse. (Lithograph from *Harper's Weekly,* March 11, 1882, 156)

Areas of fresh marsh are often characterized by a type of "trembling marsh" locally called *flotant* or *prairie tremblant,* which can adjust vertically to rising water levels through "buoyant uplift" and are distinguished by a trembling vegetative marsh mat not anchored to the ground beneath. This vegetative carpet is difficult to navigate and requires special transportation equipment, because, as one foot is pulled out of the quagmire, the other one continually sinks. This repetitive process can exhaust the uninformed and prompted some to report that "the country is so wet that we scarcely saw an acre of land upon which a settlement could be made."

Though treeless and flat, the *prairie tremblant* landscape is not without some relief. Most notable are the *chênières* (each individually labeled a *coteau* [hill] by locals) and five major exposed salt domes (Avery, Belle Isle, Côte Blanche, Jefferson, and Weeks). Salt domes, consisting of enormous, mostly buried vertical shafts of pure sodium chloride, can rise as much as 250 feet above the neighboring marshes. Although they encompass a relatively miniscule portion of the coastal plain, the salt domes boast huge economic significance through their roles as major producers of salt, petroleum, natural gas, and, at Avery Island, Tabasco hot sauce.

Increases in elevation, as seen on salt domes and *chênières*, dictate major, sometimes startling, ecological differences between contiguous parcels of land. On the coastal plain, these changes first appear in areas with a minimum elevation of eighteen to twenty-four inches, the minimum necessary to support tree growth. In the flat coastal marshes, many regional maps of which have *no* five-foot contour lines, trees are typically exceedingly rare. On ridges, on spoil banks associated with canals, and in freshwater swamps fringing the natural levees, however, trees dominate the landscape.

Source: P. Vileisis, *Discovering the Unknown Landscape: A History of American Wetlands* (Washington, DC: Island Press, 1997), 61.

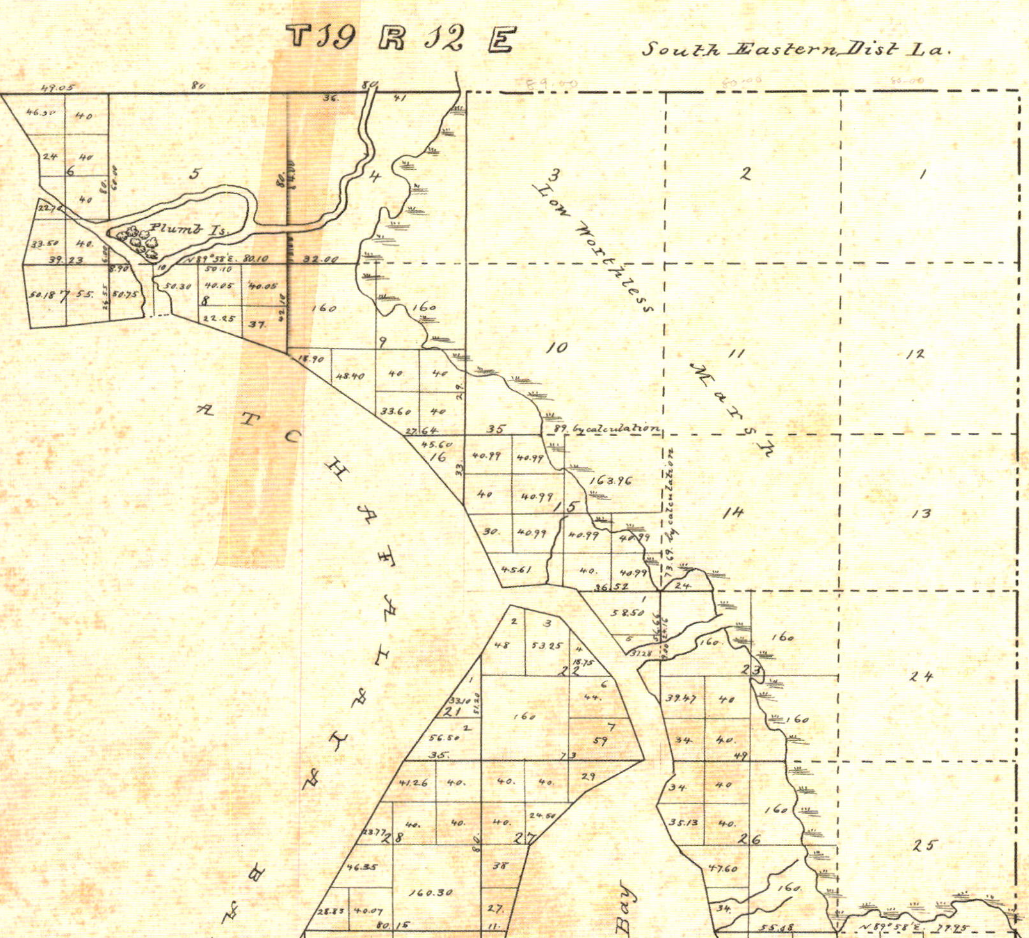

The surveyor's assessment of the coastal marshes (that is, "low worthless marsh") accurately represents popular attitudes toward the area's wetlands. (From the authors' collections, digital copy of original. The authors wish to acknowledge the assistance of the T. Baker Smith, LLC staff in locating the image)

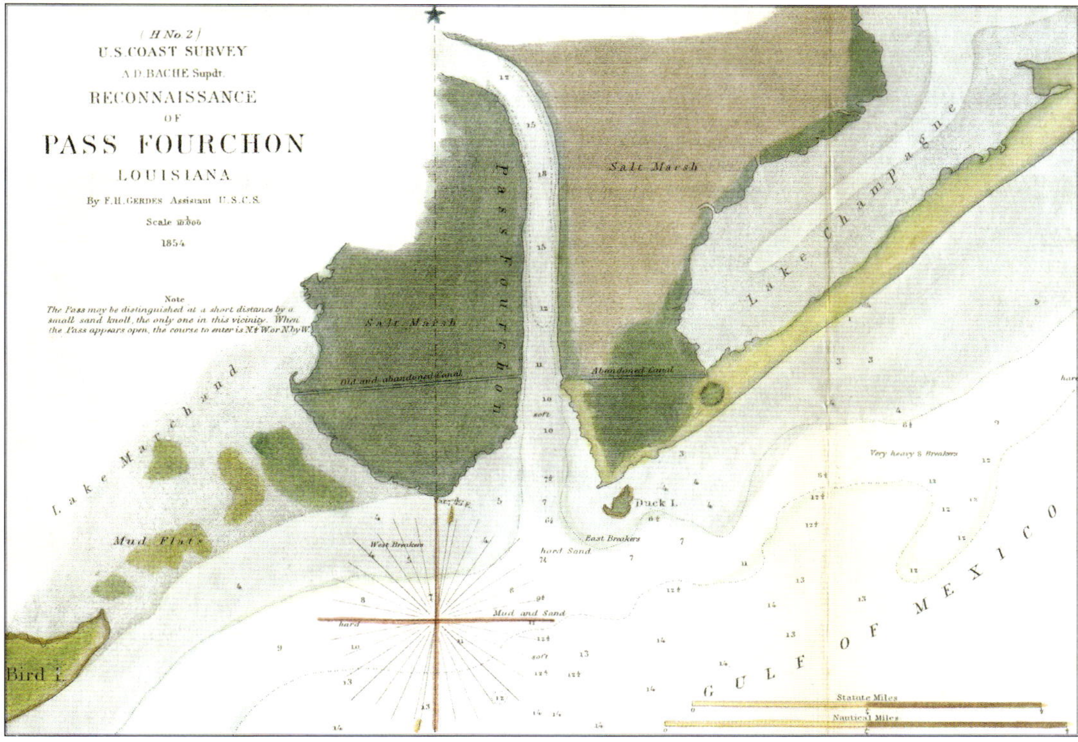

This 1854 U.S. Coast Survey map of Pass Fourchon shows the early traverse canals designed to facilitate east–west transportation in the coastal marshes. (From the authors' collections)

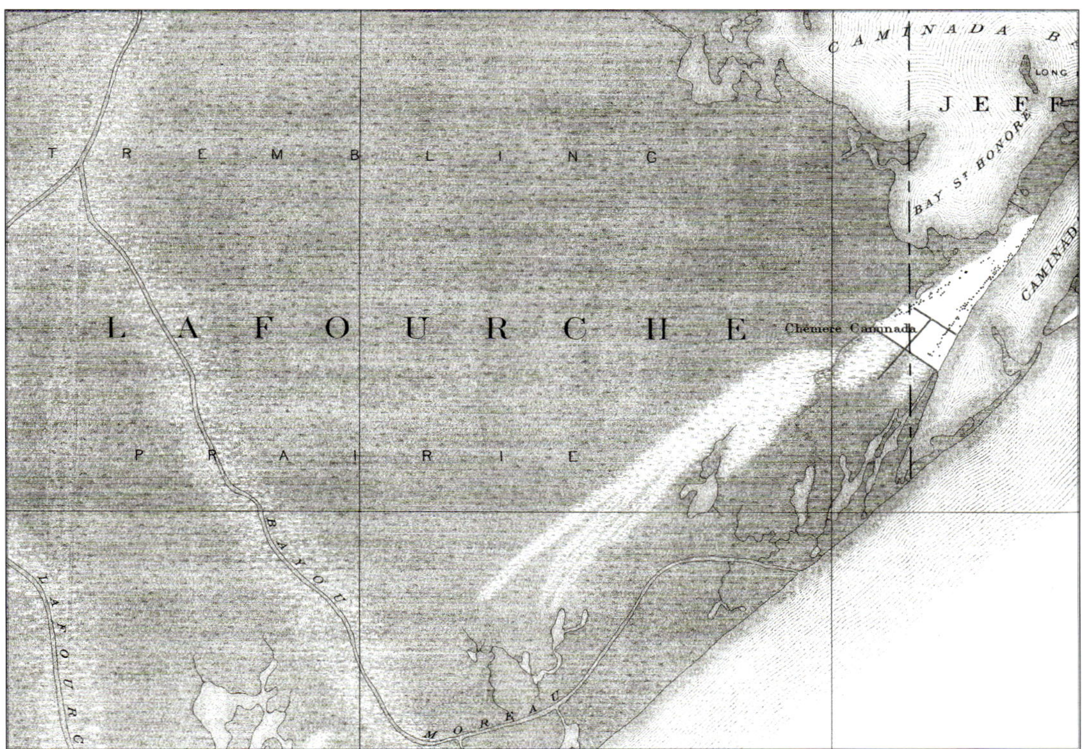

The pristine "trembling marshes" documented in this map have been largely replaced by open water in the past half-century. (From the authors' collections)

The cause of coastal erosion has been recognized for decades, but the government lacked the will to correct it. "Students of flood control are conscious of the waste of soil and plant nutrients involved in dumping into the Gulf the great load of soil carried by the Mississippi River. This waste of valuable resource may one day be prevented through a large-scale program of planned siltation."

Source: R. W. Harrison, *Alluvial Empire: A Study of State and Local Efforts toward Land Development in the Alluvial Valley of the Lower Mississippi River, Including Flood Control, Land Drainage, Land Clearing, Land Forming,* Vol. 1 (Delta Fund, 1961), 260.

Louisiana's sugarcane rows are approximately six feet wide—wider than any other row crop in North America. The workers pictured here are creating furrows in the center of the rows for "plant" cane. (Photo courtesy of the National Archives and Records Administration, College Park, MD, photographer and date unknown, call number MD, RG 54-Y, Box 1, image Y-136)

2

AGE OF AGRICULTURAL DOMINANCE

South Louisiana's topography dictated that settlers occupy the natural levees lining the region's principal waterways, and, by fortuitous coincidence, this "high ground" constituted some of the world's most fertile arable land. Although France had established Louisiana as a military outpost for the expressed purpose of encircling and containing the British colonies lining the Atlantic seaboard, it soon became evident that Louisiana would ultimately survive only if it were allowed to establish a viable economic foundation. Given the abundance and quality of the available cultivable lands concentrated on the natural levees along the colony's vital, strategic arterial waterways, agriculture eventually became the focus of Louisiana's economic development.

Maturation of lower Louisiana's agricultural economy, however, did not come easily, despite the area's superabundance of natural resources and long growing season. Most of the colony's early European immigrants had neither the skills nor the inclination to clear and till fields. This was nowhere more evident than in the first decade of Louisiana's existence (1699–1709). Because Louisiana lacked an agrarian population, the colonial garrison was completely dependent upon its lifeline to metropolitan France, but, during wartime—particularly the War of the Spanish Succession (1701–1714)—British blockaders interrupted French supply shipments for long periods, sometimes years. Consequently lacking the most basic necessities, Louisiana administrators were compelled to disband the colonial garrison on at least three occasions to permit soldiers to seek sustenance with friendly local Native American tribes. Despite these recurring crises, the garrison refused to grow its own provisions, and the colony continued on a precarious footing as long as it remained a strictly military outpost.

Louisiana's status changed significantly following its acquisition by the Company of the West (later the Company of the Indies) in 1717.

The company used the new proprietary colony as collateral to prop up the value not only of the proprietary company's stocks but also of the *Banque Royale* (Royal Bank). The *Banque Royale* maintained intimate, inextricable financial ties to the corporation through the ministrations of Scottish immigrant John Law, then the chief financial adviser to the French Crown and the chief executive officer of both the bank and the company. Law fully understood that the success of the bank and the company jointly hinged upon the transformation of their greatest shared resource—Louisiana—from a white elephant to a genuine asset. This transformation required the establishment of an economic base in Louisiana, and this, in turn, was largely dependent upon the formation of a significant civilian population and an agricultural economy to sustain it.

Efforts to achieve these goals often worked at cross-purposes. For example, to expedite European—particularly French—emigration to Louisiana, an intense advertising campaign underwritten by Law appealed primarily to gullible adventurers ready to exploit promised, but actually illusory, gems and precious metals in the lower Louisiana wilderness. These early inhabitants were supplemented by large numbers of forced emigrants, most of whom were then guests of the French penal system. Neither group was particularly well suited to the challenge of contributing positively to the rapid development of the wilderness settlement. To the contrary, these reluctant transplants—most former urbanites—promised to become a burden on Louisiana society.

Other immigrants, however, would prove to be Louisiana's salvation. Law's commercial propaganda had proven particularly appealing to Germans in the militarily contested Rhineland principalities that had, for nearly a century, been a "spoil" of war almost continuously ravaged by rival European powers. Thousands volunteered for resettlement in Louisiana, only to find themselves stranded in de facto French coastal concentration camps when company-underwritten shipping failed to materialize in a timely manner. Only a few hundred were eventually dispatched to the lower Mississippi, and illness claimed most of them during the transatlantic crossing. Although intended for settlement in present-day Arkansas, the German survivors were instead established along the Mississippi River about fifty miles upriver from New Orleans, the future colonial capital founded in 1718. On the riverbanks, the immigrants faced the daunting challenge of carving a home from the virginal hardwood forests blanketing their settlement sites. This arduous task was made infinitely more difficult by perennial seasonal flooding. With time, however, small farms emerged from the wilderness.

The truck farmers occupying the west bank of the Mississippi upriver from New Orleans—an area dubbed the German Coast in their honor—constituted the initial

Harvesting potatoes in Terrebonne Parish in the early twentieth century. The photographer reported that "twenty-five men would dig, pick and sack 260 sack cars in six days," with mules helping in the harvest. (Photo courtesy of the National Archives and Records Administration, Prints & Photographs Division, Bureau of Plant and Industry, College Park, MD, photographer unknown, January, 25, 1916, call number RG 54 Y, Box 1, Y 936 [1])

Relatively small truck farms continue to provide both New Orleans and small-town farmers markets with fresh produce. Production is generally confined to the Bayou Blue region straddling Terrebonne and Lafourche Parishes, the east- and west-bank levee lands in St. James and Plaquemines Parishes, and small farming communities north of Lake Pontchartrain. Crops include peas, cabbage, cauliflower, Brussels sprouts, mustard, parsley, spinach, onions, tomatoes, squash, garlic, and shallots. These regions have been providing fruits and vegetables for the French Market for more than a century and a half. Many of these truck farmers are under contract to provide fresh produce to New Orleans's leading restaurants.

backbone of the colony's agricultural industry, but the focus of Louisiana's continued development quickly shifted from yeomen farms to plantations in the ensuing years. This metamorphosis was predicated upon the availability of a large, cheap labor force that could be legally bound to the lands the workers occupied. Owners of Louisiana's first agricultural estates initially attempted to use indentured workers who pledged their labor for a specified period in exchange for passage to North America. (This form of labor contract later came to be identified as the credit-ticket system.) Through this new system of labor, the plantation elite treated these emigrants as colonial serfs. As indicated above, they quickly proved to be unsatisfactory workers. Owners of the *concessions* (plantations) consequently turned to Louisiana's second major source of involuntary laborers—West African slaves.

During the French period (1699–1763), Louisiana's bondsmen were drawn overwhelmingly from a common native land—Senegambia, because a Company of the Indies subsidiary, the Compagnie de Sénégal, held the French monopoly on the African slave trade. Sent to Louisiana from the Compagnie's slave detention center at Gorée Island, Senegal, most of the bondsmen were animists captured by Muslims during eighteenth-century West Africa's religious wars. A plurality of the captives hailed from present-day Mali, and, although their homeland was a semiarid area bordering the southern Sahara, most of the captives were agriculturists with experience in rice production.

The Africans and Germans would collectively play a pivotal role in establishing Louisiana's vital agricultural economy. The Germans—former peasant farmers—quickly prospered by shipping their produce to New Orleans, whose growing population constituted a virtual captive market, thanks to continuing supply shortages. The Africans, on the other hand, were responsible for the successful establishment of plantation agriculture along the Gulf Coast.

Plantation agriculture, however, was slow to become a paying proposition in lower Louisiana for a variety of reasons First, the natural levees were blanketed with old-growth hardwood forests that required many years—often decades—of back-breaking labor to clear. Second, West Africans were unfamiliar with the local climate and growing season, as well as the crops they were ordered to plant. Third, significant linguistic and cultural differences among the Africans, who were drawn from a sprawling geographic area stretching from the Senegambia region to Angola, initially made it difficult for slave owners to organize their laborers for group projects. Finally, the trauma of deracination, the horrors of the Middle Passage, arbitrary assignments frequently resulting in the breakup of extended families, malnutrition, the continuous threat of violence, and unremittingly harsh working conditions all coalesced to create a demoralized workforce that continuously sought out ways to escape bondage, often by fleeing into Louisiana's vast wetlands.

Louisiana's proprietary government had initially attempted to establish a large staple crop agricultural system built on the European model. Virtual fiefdoms were granted to French investors—many of whom were aristocrats, members of the bourgeoisie, or rural gentry. These grants stipulated that these financiers provide their own workforce consisting primarily of European indentured laborers. European workers, however, were horrified by conditions in the colony, and they returned to France at the first opportunity, causing most *concessions* to collapse by the early 1720s. The resulting void was filled by ambitious Louisiana colonists. These individuals seized upon the opportunity to acquire land and thus achieve social status unimaginable in the metropole, where landholding was the sole prerogative of the French nobility. In actuality, these budding plantation owners sought to establish feudalism in the New World—with themselves as the putative aristocracy. Because European indentured workers had proven themselves exceptionally problematic and because French forced-immigrants were an even more unacceptable option, the new landowners turned, out of necessity, to African slaves, who first entered the colony in significant numbers in 1719—one year after the establishment of New Orleans.

Despite the myriad labor problems retarding the colony's rapid transition to staple-crop agriculture in the proprietary period, an incipient plantation economy had firmly taken root on the natural levees upriver and downriver of New Orleans by the 1760s. Because mercantilism constituted the sine qua non of French economic policy, Louisiana exports were limited to items in demand—but not duplicated—in the metropole. Plantation crops were consequently determined by external market forces and bureaucratic decisions, not their suitability to Louisiana's soils, climate, or labor force. It is thus hardly surprising that Louisiana's plantation system proved less robust and lucrative than it undoubtedly would have been with fewer arbitrary restrictions. Indigo and tobacco, the colony's main French-era exports, were inferior in quality (because of weather and manufacturing issues) to their principal competitors, respectively, in Virginia and Central America. Hence Louisiana products exerted limited appeal to France, the colony's only legal and highly restricted market.

The economic fortunes of Louisiana's plantations changed little in the decades following the colony's legal transfer to Spanish rule in 1763. Spain, like France, vigorously attempted to enforce mercantilism throughout its colonial empire, and, as a result, Louisiana's chief exports remained tobacco and indigo of fair to middling quality. The colony's plantation economy consequently remained artificially constrained until circumstances forced the colony to begin diversification.

The lack of lucrative external markets was compounded in the 1790s by recurring crop failures. Although indigo was native to Louisiana and well adapted to local environmental conditions, the commercial strains of the plant—imported from the French West Indies—required dry, sandy soils to flourish. Susceptible to floods, unseasonal weather, and indigenous insects, the colonial indigo was prone to frequent crop failures. In 1794 "no indigo was made on any plantation," according to Georges-Henri-Victor Collot, a French military observer in the Mississippi Valley.

During the Spanish period, farmers—primarily immigrant yeomen, particularly Acadian exiles—began to produce small quantities of cotton, but, lacking a significant external market, production was largely limited to domestic consumption, despite the colonists' development of a pre-Whitney cotton gin. Nevertheless, a foundation was laid for rapid, exponential expansion following the United States' acquisition of Louisiana in 1803 and the regional cotton industry's access to the burgeoning textile industries in New England and, later, the United Kingdom.

Lower Louisiana's cotton experiment coincided with—and contributed directly to—the establishment of the regional sugar industry, the second major successor to the region's original staple crops. The first serious local experiments with sugar production began in the early 1750s, but various environmental and technical factors conspired to doom these pioneering attempts to failure. The sugar experimenters sought to literally and figuratively transplant the sugar industry from the West Indies, where it had already taken root, to the lower Mississippi Valley, but Louisiana's growing season was much shorter than that of the French Antilles. Louisiana's sugar crops were—and remain—highly susceptible to frosts and freezes; once exposed to freezing temperatures, sugarcane begins to die, and the plants' precious sugary sap begins to ferment. When fermentation begins, farmers have approximately twenty-four hours to get their crops to the nearest available mill. Hence weather confined the regional growing season to approximately ten months, unlike the frost-free Caribbean, where the crop typically stood in fields at least fourteen months. Longer growing seasons equate to higher sucrose content in the plant stalks. Lower sucrose content and lack of experienced sugar chemists made it impossible for Louisiana's earliest sugar growers to achieve granulation during postharvest processing, meaning that the producers could only export syrup and molasses. However, because of the colony's dearth of competent coopers, the liquid usually leaked from the crude barrels stored in hot, humid ships' holds before reaching potential markets in France.

Louisianians justifiably considered sugar cultivation a failed economic experiment, and, at most, a handful of growers continued to produce sugar for home consumption until the indigo failure of 1794 prompted New Orleans–area entrepreneur Étienne Boré to seek a more reliable crop. Because slave uprisings coinciding with, and largely resulting from, the French Revolution had disrupted the hugely profitable sugar industry in the French Antilles, the agricultural maverick opted to revisit sugar's economic potential. Boré's experimentation fortuitously coincided with the initial influx of refugees from Saint-Domingue (present-day Haiti), the sugar industry's epicenter. One of these émigrés, chemist Antoine Morin, a Boré employee, achieved what was believed to be impossible in Louisiana—sugar granulation—and a new industry was born.

Sugar, however, would not dominate lower Louisiana's economic landscape until after the War of 1812 (1812–1815). The war years witnessed an insect infestation that devastated the then-ascendant regional cotton crop. Furthermore, a British blockade of Louisiana ports prevented export of the shrinking harvest. Desperate for a reversal of fortune, many planters on the natural levees transitioned from cotton to sugar production in the decade following the Battle of New Orleans (January 8, 1815), the war's climactic concluding act. Commenting on the transition, the geographer William Darby noted that "an acre of sugar cane will, in ordinary seasons, produce more than 1000 pounds of sugar; which, at a moderate price, will amount to more clear profit than any other product yet cultivated in Louisiana." By 1850, when the first decennial agricultural census of the United States was compiled, parishes in Louisiana's coastal plain produced 93.36 percent of the Pelican State's sugar and 89.42 percent of its molasses output. However, the conversion to sugar was not universal, and cotton would remain an important crop on the natural levees and, eventually, in various coastal wetland communities. Indeed, cargo manifests drawn at five-year intervals from steamboats servicing Bayous Teche, Courtableau, Lafourche, and Vermilion between 1825 and 1860 indicate that the western Sugar Bowl parishes shipped at least 126,599 bales of cotton to New Orleans, the nation's premier cotton market during the regional sugar industry's antebellum ascendancy.

The sugar industry's dominance in the early nineteenth century had myriad economic, social, cultural, and environmental ramifications for the Louisiana coastal plain. Because of its relatively high profit margins, the emerging sugar industry served as a magnet for well-heeled tidewater Virginia and Carolina families seeking fresher lands and promising opportunities in emerging American agricultural markets. These immigrants and their established Louisiana neighbors made enormous capital investments in the establishment of the infrastructure necessary for the industry to thrive. These speculative ventures initially focused upon land acquisitions necessary to sustain extensive monocrop production, the huge manpower demanded by this very labor-intensive industry, and the milling equipment required to transform the raw product (unprocessed cane) into value-added commodities (syrup, molasses, and raw sugar). Within two generations, the natural levees were transformed from a geographic province of small-to-moderate-size farmsteads to sprawling plantations; the formerly Catholic, white population became overwhelmingly African American, as English-speaking Protestant bondsmen were imported from the surplus slave populations along the eastern seaboard; and enormous swaths of old-growth woodlands were denuded first to clear fields and, later, to sustain wood-fired mills. The 1850 US agricultural census indicates that coastal plain residents had cleared and exploited 664,816 acres (1,038 square miles—an area slightly smaller than Rhode Island) in some of the world's most forbidding and pestilential terrain. It is thus hardly surprising that during the decade before the Civil War, South Louisiana sugar plantations were importing coal to fuel their milling operations, and they continued to do so until the second decade of the twentieth century.

Despite the high perennial cost of operation, sugar planters enjoyed huge incomes and lavish lifestyles—at the cost of untold human suffering by their slave labor force. Nottoway, the largest surviving southern plantation home (built near White Castle by a Virginia immigrant family), is an enduring monument to the sugar industry's golden era.

However, the Louisiana sugar industry's halcyon days were numbered. The Civil War (1861–1865) ushered in a new, catastrophic chapter in the Sugar Bowl's history (see Table 1). The Union blockade of the Mississippi River (both upstream and off the Louisiana coast) denied Pelican State producers access to crucial North American and international markets, forcing sugarcane growers to drastically curtail production. In addition, the Union occupation of New Orleans in the spring of 1862 immediately created a debilitating labor crisis. Although the South Louisiana sugar-growing parishes were specifically exempted from the Emancipation Proclamation (January 1, 1863), the Union occupation brought an effective end to slavery in the Sugar Bowl. By the spring of 1863 thousands of slaves in the permanently occupied areas around the Crescent City were, according to one contemporary observer, "averse to doing any kind of work" for their erstwhile masters. Thousands of additional slaves deserted their rural plantations for New Orleans–area "contraband" camps supervised by the Union army after every Northern military incursion into Louisiana's interior.

Table 1
Louisiana sugar production

Year	Tons
1861	269,000
1862	51,000
1863	42,500
1864	5,400

One scholar has estimated that South Louisiana's manual workforce may have constricted by as much as 50 percent in the early postbellum era. The resulting labor crisis, which persisted well into the postwar era, was compounded by various man-made and natural disasters. Fields that lay idle for most of the war required clearing, but manpower and draft animals were difficult and expensive to procure. In addition, the region's transportation infrastructure sustained severe wartime damage and was virtually unserviceable by 1865. These almost insurmountable challenges were exacerbated still further by the dearth of available credit. The few state banks that survived the war were understandably cautious about dispensing credit—the lifeblood of modern agriculture—because of the heightened risks that postwar lending entailed. These fears were amply justified, for continuing labor issues, insect infestations, highly destructive vernal floods, and deadly yellow fever epidemics devastated the local workforce in the weeks prior to the annual sugar harvests. These postwar issues ensured that the surviving sugar operations would fail. For three consecutive years following the war's conclusion, plantation-based sugar production declined sharply.

After three recurring crop failures, banks refused to extend additional credit to now overextended farmers, but because bankruptcy was so pervasive, financial institutions were reluctant to call in their debts until the national panic (depression) of 1873 forced their hand. Regional courthouse records show an unprecedented spike in public auctions of farms and plantations in the sugar parishes in 1873—a reflection of how unforgiving Reconstruction-era life had become, even for families of means. The result was a regional economic crisis that endured for decades—arguably until the beginning of World War II—and the rapid transformation of the area's formerly stratified society into one closely resembling a modern Third World model: atop the local hierarchical pyramid, wealth was overwhelmingly concentrated in the hands of a small elite. A tiny middle class composed of urban professionals and surviving small landholding interests occupied the middle tier, and former slaves and dispossessed yeomen—all now reduced to tenancy—constituted an enormous underclass.

The waxing and waning fortunes of the South Louisiana sugar industry had an incalculable impact upon the economic development of the entire coastal region. The rise of plantations and the corresponding increase in property values forced poor whites, most of whom feared debt, to occupy increasingly marginal settlement sites. East of the Atchafalaya River, these economic refugees first occupied ridges of relatively high land in the swamps behind—and between—the colonial settlement sites, then the progressively narrower and lower natural levees in the coastal

Between 1929 and 1938, the freight boat *Margie* made regular trips from Lake Arthur to Grand Chenier to deliver a wide assortment of goods, including a new car that could only reach this isolated coastal community by boat. (From the authors' collections, used with permission of Norman McCall)

Isle de Jean Charles (better known locally as Isle à Jean Charles) is a narrow ridge situated in lower Terrebonne Parish's coastal marsh. The ridge is the historic homeland and burial ground of the Isle de Jean Charles band of Biloxi-Chitimacha-Choctaw Indians. The community and its associated artifacts are at risk from subsidence and rising water levels. (Photo by the authors, 2007)

marshes. West of the Atchafalaya, migrants occupied prairie lands and the *chênières* in the coastal marshes.

All of these marginal sites were deemed undesirable by the mainstream agricultural community because they were ill-suited to extensive monocrop agriculture. The prairies boasted an underlying claypan impenetrable to the era's wooden farming implements, while the *chênières* and lower natural levees, which were highly vulnerable to either (or both) seasonal and storm-driven inundations, had limited arable land and, other than a few navigable waterways, no functional transportation system. Despite the inherent environmental constraints and physical isolation of these remote settlement sites, the economic refugees consistently engaged in subsidence or near subsistence agricultural activities following their relocation.

The corpus of scholarly literature on migrants indicates quite clearly that emigrants consciously and subconsciously seek out the familiar in their adopted homes. This is particularly true of their acquired economic expertise. In the prairies, settlers engaged in subsistence agriculture and ranching. Since southwest Louisiana's prairies possessed almost ideal conditions for development of a livestock industry, the area became one of America's oldest and largest cattle-producing areas. Local pioneers quickly established herds on the open range prairies west of upper Bayou Teche and the Vermilion River. In 1769 envoys from Spanish governor Alejandro O'Reilly discovered more than 2,000 head of branded cattle at the Opelousas Post (all or parts of present-day St. Landry, Evangeline, Acadia, Cameron, Calcasieu, Allen, Jefferson Davis, and Beauregard Parishes) and approximately 1,000 head at the Attakapas Post (encompassing modern-day St. Martin, St. Mary, Lafayette, Vermilion, and Iberia Parishes). By 1803, when Louisiana became a US territory, branded cattle in the Opelousas District alone numbered over 50,000 head.

Ranchers' use of the open prairies, where cattle coalesced into massive unsupervised herds, dictated the need for strict branding regulations to facilitate recovery of errant stock and to mitigate the ever-present threat of rustling and predation by local criminals. Governmental authorities scrupulously maintained registries of brands, which effectively became business trademarks, from the very earliest days of prairie colonization. The brand book of the Attakapas and Opelousas Districts, for example, records more than 2,700 brands registered between 1760 and 1888 by *vachers* (cowboys) with French, Spanish, Acadian, and Anglo-Saxon surnames.

As the mélange of registered surnames suggests, the prairie ranching area was a regional cultural melting pot, with each new immigrant group—Hispanic, Creole of Color, and Anglo—contributing its own ranching traditions to those of the founding

Cattle were often trailed from ranches west of the Sabine River, across the prairie, through the Atchafalaya Basin, and onto steamboats to be slaughtered in New Orleans. (Lithograph from *Frank Leslie's Illustrated Newspaper*, June 16, 1883, 272)

"At great intervals on the prairies, homesteads had been established by ranchers who were engaged in raising cattle and horses. Large herds of these animals wandered at will, as there were no fences. . . . These prairies were covered with a luxuriant growth of grass, most of which was burned off by the cattlemen in late summer and early winter, so that the livestock would have short, green grass on which to feed during the winter. This was southwestern Louisiana as I knew it as a boy."

Source: E. A. McIlhenny, "Major Changes in the Bird Life of Southern Louisiana during Sixty Years," *Auk* (1943): 541–542.

Acadian and French families. The impact of Spanish-speaking and Creole of Color settlers was especially notable in the eastern and upper prairies, respectively, while Anglo influences were particularly pronounced in the *Chênière* Plain.

The prairie region's evolving ranching traditions did not place a premium on selective breeding or other forms of stock improvement. In 1803 French traveler C. C. Robin noted that introduction of external stock was "an activity much neglected in Louisiana," and the region's original strain—evidently longhorns smuggled into the region from Spanish Texas in the eighteenth century—easily dominated the regional cattle industry for more than a century and a half.

Although locals devoted precious little attention to the improvement of native stock before the 1880s, the Louisiana coastal plain witnessed one notable effort at scientific breeding. American cattlemen introduced the Brahman (*Bos indicus*) variety into the United States in 1849, and Richard Barrow, a plantation owner near St. Francisville, imported the strain into lower Louisiana in 1854. By 1866 the breed was established also in Texas, where extensive tracts of salt grass provided pasturage. Brahmans, originally from the Indian subcontinent, flourished in the Gulf coastal plain's hot, humid climate, and their natural resistance to the area's indigenous pests made them far superior to other introduced breeds—particularly those from northern Europe. The Brahman also produced meat far superior in quality to that of the area's now dangerously inbred native stock. As a result, Brahmans became coastal ranchers' breed of choice by the early twentieth century. (Over succeeding generations, enterprising ranchers have developed a variety of crossbred strains—most notably the Brangus—that combine the hardiness and disease resistance of the Brahman variety with breeds boasting superior beef-to-bone ratios.)

By the late nineteenth century, though other agricultural enterprises were beginning to compete with husbandry, the cattle industry continued to flourish. In the last decade of the century, cattle were being shipped to many parts of the United States and to Cuba from the state's marshes and near sea-level reclamation sites. However, the southwestern Louisiana ranching tradition, once centered in the inland prairie, is now concentrated in the reclamation zones along the northern fringe of the marshes.

Throughout coastal Louisiana, early efforts to reclaim large tracts of Louisiana's wetlands were largely unsuccessful—with the notable exception of tracts designed for use by livestock. As immigrants occupied the open prairies and rapidly converted them to agriculture, native ranching families began to drain and reclaim marsh areas for use as pasturage. In the Deltaic Plain, these reclamation projects were generally associated with natural levee lands. On the *Chênière* Plain south of Louisiana Highway

14 (perhaps more aptly dubbed the Reclamation Road), cattlemen allowed their herds to graze the marsh, utilizing pumps to create vital freshwater impoundments.

In both regions, weather and indigenous pests dictated the seasonal movement of herds. In the cool, relatively dry month of October, local drovers began moving herds from summer pastures to the coastal wetlands. Although cattlemen in both areas sometimes utilized cattle boats constructed by local boat builders to transport their herds from one grazing site to another, ranchers more typically moved their herds by means of cattle drives over treacherous overland routes in the nineteenth and early twentieth centuries. Cowhands sloshed through knee-deep mud by day and coped with numberless mosquitoes and other pests at night. A small vessel—the chuck wagon of the coastal country—met the herd at designated places along water bodies paralleling the marshland trails. Manned by one or two men, each functioning as both sailor and cook, these vessels sometimes provided sleeping quarters for weary cowhands. After a herd had safely reached marshland pastures, the cowhands loaded their saddles aboard the boat and returned to the ranch—usually a two-to-three-day trek.

Maturation of Louisiana's canning industry propelled the expansion of Louisiana's fishing industry, and small shipyards, like this one in Lake Arthur, were kept busy building wooden-hulled ships. (From the authors' collections, used with permission of Norman McCall)

In the marshes, cattle roamed the open range at will, eventually coalescing with cattle from other ranches at Constance Bayou, Dan's Ridge, Eugene Island, Long Island, Oak Grove Ridge, Pecan Island, Chenier Perdue, Little Chenier, Creole, and Chênière au Tigre to form massive herds—upward of 6,000 head. Rather than sort the stock by brand—a nearly impossible task in the marsh—when they returned in the spring, the stockmen drove the communal herd along the Gulf beaches to designated points near Vermilion Bay approximately eighty miles to the east. At the coastal trail's eastern terminuses, ranchers placed their cattle on sternwheelers or barges for shipment to unloading facilities in the Vermilion Parish communities of Boston, Abbeville, or Intracoastal City (an important logistic support base for the offshore oil business that was formerly a small community for fishermen, trappers, and farmers).

...on a SLOW BOAT to PECAN ISLAND...

It's a long trip, the voyage to one of Louisiana's offshore outposts, but with a sense of humor and a camera, time passes quickly. Magazine's "Wanderin' Willie" sends in characteristic memo of journey

A rare article about lower Vermilion Parish. Notice the comment that Pecan Island was an "offshore outpost." (From the authors' collections, used with permission of Norman McCall)

Once the railroad was completed across Louisiana's prairie region in 1880, coastal plain ranchers abandoned long cattle drives. Since no paved or shell roads existed in the marshes, steamboats and other watercraft moved the animals to the nearest rail shipping point. In the spring of 1899, one shipment of marsh cattle included 1,900 head.

The *Chênière* Plain's transition to ranching paralleled the metamorphosis of other wetlands communities as they converted from subsistence agriculture to commercial fishing. Even communities now universally associated with commercial fishing, like Pierre Part along the Atchafalaya Basin's eastern periphery, initially cultivated small

After property and ownerships issues were addressed by state and district courts and property titles were secured, many old cattle trails had to be replaced with new routes capable of handling the large herds that were a by-product of better range management practices. Each October and April herds were trailed or trucked, sometimes as far as 70 to 100 miles. This practice permitted the cattle to escape the summer mosquitoes and to take advantage of the marsh's excellent winter range. In fact, some ranchers once wintered their herds on Marsh Island or Last Island.

Shortly after World War II, ranchers introduced truck transportation, and multiday cattle drives ended. Small-scale drives, however, continue, and drovers use dogs and whips to corral cattle for shipment to an appropriate loading facility. This herding activity is one of the country's last horseback trail drives. No longer do the animals go to railheads, but into waiting eighteen-wheelers for the ride to upland pastures.

Eventually, large landowners controlled the range, so traditions had to change, as there were no longer unfenced pastures. Out of necessity herds became smaller. Even so, ranching continues as an established part of the local culture. Horses, cowboy hats, boots, cattle, bridles, ropes, belt buckles as large as small plates, and jeans are as much a part of the cultural heritage of the *Chênière* Plain as they are a part of western grazing tradition. Like the American and Canadian West, "cowboying" on the *chênières* has not changed too much from its beginning. Cattle drives are not as important as they were in the early days of the industry when the cattle drives were so large that children were excused from school for a week in the spring to help gather, brand—often up to 1,000 calves a day—and trail cattle from the marsh to sites around the marsh fringe communities of Ged, Edgerly, Starks–Big Woods, Kinney, Lawton, Vincent, Vinton, and Toomey. Cowpunchers used boats occasionally to move their herds, so a "trail drive" regularly involved floating the cattle to market. With from 15,000 to 20,000 head grazing north of Johnson Bayou, along with up to 100 horses, these drives to summer pastures were impressive events. With 60,000 hooves, 30,000 horns, and about fifteen million pounds of force, it took a great deal of skill to keep this mass walking in the same direction.

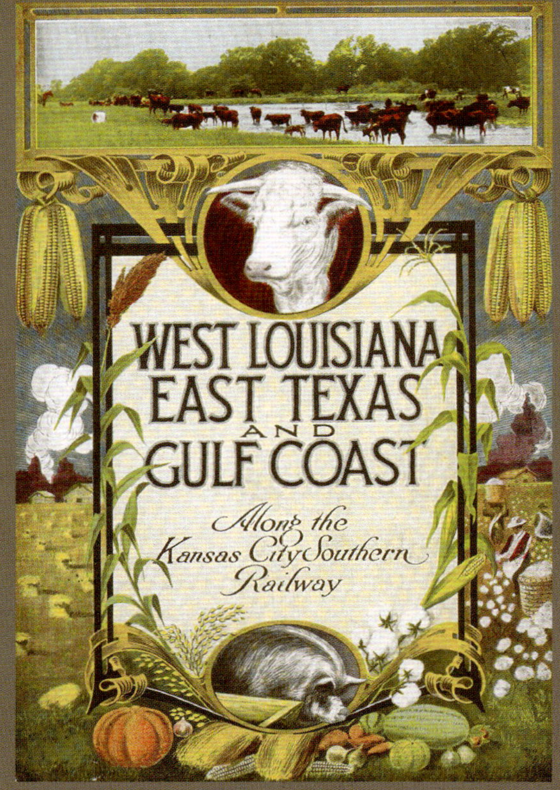

In order to attract settlers, coastal Louisiana land developers, including railroad companies, produced sophisticated marketing booklets for distribution to potential transplants. (Cover photo from *West Louisiana, East Texas, and the Gulf Coast along the Kansas City Railway*, 1917, Kansas City Southern Railroad, courtesy of the Beinecke Library, Yale University)

Two important technological innovations have had a direct influence on livestock production's dramatic gains in the post–World War II era. First, animal husbandry and scientific breeding applications perfected in the latter part of the nineteenth century helped to develop an animal stock that is particularly well suited to the region's ecological conditions. Second, reclamation techniques in the latter part of the nineteenth century and the first two decades of the twentieth century converted unused wetlands (also known at the time as wastelands) to prime pasturage. Also, cattlemen use rice fields for grazing in rotation with their marsh pastures to add weight and improve their livestock. This is an effective management practice and reduces some of the problems associated with overgrazing. Seasonal grazing of the marshland, along with rice, is more profitable that rotating rice farming with cattle. This is due, in part, to the limited cost associated with marsh grazing.

plots for home consumption and limited export to urban markets, while engaging in small-scale cattle production. They were, in a word, farming communities.

Agricultural dominance in the coastal plain came to an abrupt end during the early postbellum era as a result of recurring major floods on the scale of the benchmark 1882, 1912, 1913, 1916, and 1927 inundations. As in the plantation belt to the north, these marginal agricultural communities faced a harsh postwar economic imperative—adapt or die.

In the plantation-dominated Sugar Bowl area along the natural levees, farmers adapted by first experimenting with immigrant laborers—particularly Chinese "coolies"—and then by establishing tenantry and the crop-lien system. According to the 1900 census of Louisiana, 89.1 percent of all Louisiana farmers were tenants. Residents of fringe communities, on the other hand, turned to commercial fishing, trapping, and hunting. However, the success of these new commercial enterprises, particularly hunting and fishing, was contingent upon a paradigm shift in transportation technology—a technological revolution that would paradoxically facilitate the passing of one agricultural era in the wetlands, while concurrently giving birth to another.

The role of women in coastal Louisiana trapping is now largely forgotten, despite their critical importance to the industry. These trappers were based at Chênière au Tigre in lower Vermilion Parish. (Photo reprinted from *History of Vermilion Parish, Louisiana* [1983], courtesy of the Vermilion Historical Society)

Prior to the late 1850s, commercial agriculture in Louisiana's coastal plain was entirely dependent upon waterborne transportation—particularly sail-powered vessels and steamboats—for market access. Waterborne transportation, however, was risky and rather undependable. Interior waterways connecting the Louisiana coastal plain's agricultural parishes with the Crescent City were literally lined with the hulks of sunken boats claimed by snags, fires, or burst boilers. In addition, many of the region's vital watery thoroughfares were only seasonably navigable. Farmers and planters in many parishes, particularly those west of the Atchafalaya, had virtually no access to New Orleans during the "low-water season," which commonly extended from July to December.

Even so, market accessibility improved drastically shortly before the Civil War. In 1857 the New Orleans, Opelousas, and Great Western Railroad connected Brashear City (now Morgan City) with Algiers (opposite New Orleans), and, while rail transportation remained costly, it was dependable and fast. Speed was absolutely crucial to the development of emerging postwar industries. Fish and wild game taken respectively by commercial fishermen and hunters were highly perishable and had to be delivered within hours to New Orleans markets. (Development of these industries will be examined more thoroughly in the next chapter.)

The technology that enabled some coastal plain residents to abandon farming ironically permitted other South Louisianians to enter the region's agricultural markets.

The Union occupation of New Orleans in the spring of 1862 immediately disrupted the underpinnings of involuntary servitude in the plantation belt. Throughout the remainder of the war and Reconstruction, there was a continuous struggle between socioeconomic and political forces bent upon reestablishment of involuntary servitude, on the one hand, and a free market labor economy, on the other. The outcome was almost continuous labor strife. Rather than resolve the festering problems with the local labor force—now consisting of resentful former slaves and disaffected white tenants—Pelican State planters attempted to simply bypass such issues as wages, housing, and working conditions by importing new, and from the employers' perspective, hopefully more tractable European laborers.

The historian William Ivy Hair observed that after "the Civil War many Louisiana planters had incessantly complained about the general 'worthlessness' of African American labor. This topic seemed to rank alongside politics and the weather as a favorite subject of genteel conversation." However, the Pelican State's premechanized monocrop industries remained totally dependent upon the availability of abundant, cheap, obedient, and relatively compliant labor. Cognizant that their counterparts in the Caribbean rim "had negotiated a relatively smooth transition from black slavery to freedom by supplementing" their black laborers with indentured immigrants, sugar parish planters hoped to emulate their success.

In the early 1870s Iberville Parish sugar growers briefly experimented with Swedish and, later, German workers, but these northern Europeans refused to be "treated like negroes." Enterprising planters then turned their attention to Chinese "coolies," who were recruited through San Francisco labor brokers, in a second attempt to implement the Caribbean model. Shortly before the American Civil War, Cuban growers had successfully addressed their own labor shortages by importing Chinese laborers, whom they often "coerced into semi-indentured servitude." In Louisiana, however, the Asians proved far less docile than anticipated, and, again, the experiment proved short-lived. (The failed experiment nevertheless had an enormous inadvertent impact upon the development of the state's shrimp industry.) Subsequent attempts by Louisianians to press Italians/Sicilians into service as cane field laborers proved equally unsuccessful.

When Old World laborers provided unsatisfactory, the region's power brokers devised an innovative solution to South Louisiana's seemingly insoluble postbellum labor crisis and its attendant economic woes—recruitment of white northern transplants. Northerners offered Louisiana communities numerous advantages over their immigrant counterparts. First, northern transplants absorbed their own

Following failed post-Reconstruction-era attempts to integrate them into the sugar industry workforce, Italian/Sicilian immigrants congregated in the greater New Orleans area, becoming the backbone of Louisiana's strawberry industry, centered in Tangipahoa Parish north-northwest of New Orleans. Through crop experimentation beginning in 1866, residents of Tangipahoa Parish's coastal plain discovered that the local soil and climate were ideally suited for strawberry cultivation. (Area horticulturists insist that local soils give cultivated varieties a unique flavor.) Production soon increased to the point that the local division of the Illinois Central Railroad introduced express service to northern urban markets. By 1900 trainloads of berries were shipped north daily during the peak season (March–May). In addition, an influx of migrant "pickers" became part of the local cultural landscape during the harvest.

Since 1922 work by Louisiana State University research scientists helped develop new high-yield varieties that bear berries early in the season, giving Louisiana growers a crucial leg-up on the competition in the highly lucrative northern and eastern markets. In the early 1940s, more than 2,500 farms encompassing about 5,200 acres produced more than seven million quarts of these improved varieties. By the end of the twentieth century, however, the number of farms involved in strawberry production and the acreage devoted to this crop had declined sharply. In recent years, fewer than 100 commercial and "backyard" growers remain in the business. Yet strawberries remain the state's most important fruit crop, with Tangipahoa Parish generating $11.5 million in gross revenue during the 2010 season.

Since the early twentieth century, migrant workers—like this transient harvester—have been a crucial component of Louisiana's strawberry industry. (Photo courtesy of the National Archives and Records Administration, Prints & Photographs Division, College Park, MD, photographer Russell Lee, 1938, call number LC-USF33-011842-M3)

In order to meet their labor demands, early twentieth-century truck farmers hired transient workers and their children as young as four years of age as day laborers. (Photo courtesy of the National Archives and Records Administration, Prints & Photographs Division, Department of Commerce and Labor, Children's Bureau, 1912–1913, College Park, MD, photographer Lewis Hine, June 7, 1909, call number NWDNS-102-LH-829)

transportation costs. Second, they generally shared the economic, political, and social values of South Louisiana's elite. Third, and perhaps most important during the heyday of white supremacy, prominent Louisianians relied on the newcomers to "whiten" regional populations dominated by African American plantation workers. Finally, the transplants' presumed economic desperation and reputed industriousness made them perfect candidates to develop marginal lands, especially prairie and reclaimed marshlands traditionally shunned by native farmers, but heavily promoted by enterprising real estate speculators. By the turn of the twentieth century, even Louisiana's most unabashed southern chauvinists—such as US senator Joseph E. Ransdell—sang the praises of "Yankee 'energy.'" In 1908, for example, the *Crowley Signal* solemnly proclaimed that "we want settlers who are ambitious to own a strip of soil and to become citizens and respecters of the law, builders of schools and churches, and makers of a commonwealth."

Enlistment of northerners, however, posed a challenge because of the notorious deficiencies of publicly funded immigration bureaus. On March 17, 1866, the state of Louisiana established a recruitment office in New Orleans, but plagued by chronic legislative underfunding and the politicization of its administration, the agency failed to have a significant impact upon immigration during Reconstruction (1862–1877) and beyond. In fact, the agency was so ineffective that, by the early 1880s, several southern Louisiana parishes—including St. Landry, Lafayette, Vermilion, and Orleans—felt compelled to establish their own bureaus. By the late nineteenth century, however, these regional bureaus—also chronically underfunded—were complemented by well-heeled private interest groups representing wealthy land speculators, railroad companies, and venture capital corporations. These boosters targeted with great success midwesterners seeking to escape the region's harsh winters. Indeed, Louisiana's population surged from 1,118,588 in 1890 to 1,656,388 in 1910—a 48 percent increase—in part because of the influx of the recruited labor force.

The midwestern immigrants, supplemented by refugees from the Prussianization of a unified Germany, occupied huge swaths of the prairie lands between Lafayette and Lake Charles. Additional midwesterners—highly susceptible to recruiters' propaganda claims—eagerly bought up properties in reclamation projects in the upper coastal marshes bordering the southern prairies. The influx was so great that regional property values reportedly increased 120 percent between 1880 and 1900.

The transplants immediately transformed agriculture in their adopted homeland. Unlike their counterparts on the natural levees where single-crop agriculture had dominated the economic landscape for decades, prairie residents—who focused

At Avoca Island, pump drainage allowed property owners to farm reclaimed land. (Photo courtesy of the Morgan City Archives, Morgan City, LA, date unknown)

The quote that follows is typical of the highly effective promotional literature designed to attract the attention—and dollars—of farmers seeking cheap lands with which to make a new start: "With suitable drainage there is scarcely a foot of soil within the limits of Cameron that could not be made to yield abundantly. At present not over one-fourth of the lands capable of cultivation without reclamation are being utilized. . . . There are . . . a large amount of State lands still vacant, some of which might be cultivated without reclamation, but the greater part being swamp or coast mash, will require expensive improvements to render it available for agricultural purposes."

Careful observers soon realized, however, that these reclamation projects were not only unsustainable but that they also posed a long-term environmental hazard: "While the building of levees, the clearing of lands, farming, the building of cities and accompanying drainage projects have been justified in order to meet the needs of man, let us inquire whether it is not possible to overdo this civilization process? I say, emphatically, yes, if we expect to rely on any of the products of our wet lands as sources of food supply, wealth, and recreation. Even the real estate promoters who are making the gulf coast [*sic*] 'America's Newest Playground,' while extolling to the highest the natural wealth of this area, do not yet realize that they are 'killing the goose that laid the golden eggs.'"

Sources: W. H. Harris, *Louisiana Products, Resources, and Attractions: With a Sketch of the Parishes: A Hand Book of Reliable Information Concerning the State* (New Orleans: New Orleans Democrat, 1881), 121–123. P. Viosca, "Louisiana Wet Lands and the Value of Their Wild Life and Fishery Resources," *Ecology* 9(2) 1928): 227.

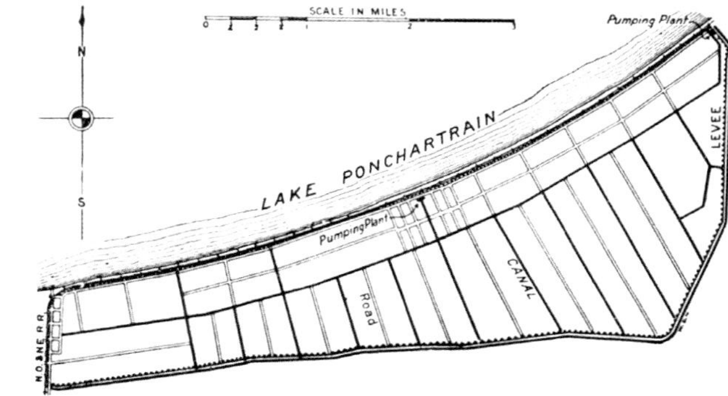

Fig. 5.—Sketch map of New Orleans Lake Shore Land Company tract.

The New Orleans Lake Shore Land Company (often identified as the Northern States Citrus and Realty Company) began planning to reclaim land adjacent to Lake Pontchartrain in the early part of the twentieth century. After the hurricane of 1915 and the severe freeze of 1917, the company went bankrupt, and the proposed community never materialized. Samuel Zemurray, president of the United Fruit Company, purchased the tract in 1934. Twenty years later, the property was sold to Samuel Brown, a Las Vegas casino owner. In 1959 the LaKratt Corporation acquired the land and subsequently transferred title to Lake Forest, Inc. The latter company eventually drained and developed the tract. The development is often called Lake Forest. (From C. W. Okey, *The Wet Lands of Southern Louisiana and Their Drainage*, Bulletin No. 71 [Washington, DC: US Department of Agriculture, 1918], 27)

Acadia Parish's German immigrants complemented the technological field operation improvements introduced by midwestern settlers by introducing levees and irrigation systems to inundate rice fields. Rice farmers throughout Louisiana quickly adopted their irrigation methods, and, for decades, private and semiprivate companies managed extensive gridlike canal networks. These networks subsequently facilitated saltwater intrusion into the interior marshes.

primarily upon ranching—were generally far behind the curve in adopting new agricultural technology. The immigrants, on the other hand, brought with them state-of-the-art field implements and steam-powered processing equipment. They first attempted to utilize this equipment in the production of wheat, the crop they were most accustomed to cultivating. However, wheat was ill-suited to the Gulf Coast climate, and these frustrated farmers were compelled to turn quickly to the most compatible local substitute—rice.

Rice had been produced in Louisiana since colonial times—initially to feed West African slaves and, later, as a fallback crop whenever the locally dominant cereal crop—corn—failed. On the prairies, residents produced "providence rice," cultivated in shallow depressions (known locally as *platins*), in which production was entirely dependent upon providential rainfall. The newcomers, however, quickly fenced off the western prairies' open range and instituted monocrop agriculture with the assistance of venture capitalists who established the necessary irrigation infrastructure.

The landscape's metamorphosis was immediate and dramatic. In 1860 the coastal plain parishes produced 99.27 percent of Louisiana's rice harvest. Three southeastern parishes—Plaquemines, Lafourche, and St. Charles—accounted for 92.22 percent of the state total. Following the arrival of the midwesterners, however, the industry's locus shifted immediately to the southwestern prairies, where it has endured for more than 150 years.

Statewide acreage devoted to rice production increased dramatically, with individual parish outputs rising between 342.6 and 7,663.4 percent between 1889 and 1899. By the turn of the twentieth century, Acadia, Calcasieu, Cameron, and Vermilion Parishes collectively accounted for 73 percent of both the state's rice acreage and output.

The rice industry's impact upon the state's overall economy was equally dramatic. In 1860 South Carolina and Georgia together produced more than 90 percent of the national rice crop. Louisiana accounted for only 3 percent of America's total output. By 1889, however, Louisiana's coastal lowlands were the nation's largest rice producer, accounting for 58.83 percent of the US rice harvest, nearly twice the output of South Carolina and Georgia combined. A decade later, Louisiana produced nearly 70 percent of the national total.

Despite the robust growth of the Louisiana rice industry, all was not well in the coastal region's agricultural sector, for the dawn of a new century presented local farmers with a fresh set of challenges. The American acquisition of rival rice- and sugar-producing territories—Hawaii, Cuba, Puerto Rico, and the Philippines—in 1898 immediately jeopardized lower Louisiana's former competitive advantage with

By 1892, 100,000 tons of Acadiana rice were shipped by the Southern Pacific Railroad—10,000 railroad carloads. Rice mills were being constructed, a system of primary and secondary irrigation canals were excavated, and rice fields were expanded, with Vermilion Parish having the state's largest network of irrigation canals. During the harvest, entire communities would work together to bring rice bundles from the field to the nearest threshing machines. The rice was then separated from the straw and sacked. Mule- or horse-drawn wagons made continuous trips from the fields to the threshing machines.

Even though Louisiana in the late nineteenth century was, along with Texas, the backbone of the rice industry, harvesting rice was a labor-intensive business. The rice plant was cut and tied by a machine called a binder into ten-inch-diameter bundles. A two-to-three-man crew operated the binder. Small bundles or shocks (rice stacked upright in a field) were assembled in the field. If the weather was dry, these pyramidal stacks were allowed to dehydrate in place. A crew of up to twenty men transported the bundles to a central point and separated the grain from the straw with a steam-powered thresher. The harvested grain was placed in burlap bags for transport, drying, and storage. With development of a combine, most of this job could be completed with a very small crew of farmhands. Grain could be stored in a bin on the combine and periodically conveyed to a self-unloading wagon pulled by a tractor, which in turn hauled the rice to a highway where trucks transported the product to a regional milling and processing facility.

All of this agricultural development led to the demise of the tall-grass prairies' large-scale ranching tradition, with the exception of thousands of acres in and along the northern fringes of the coastal *marais*. In the late 1800s ranching operations in these areas exported cattle that grazed the marsh. As the farms encroached upon cattle country, newly established farmers employed range hands as "bird miners." Originally, they would ride a horse around the rice levees cracking their whips to scare off the blackbirds. Later, when they could afford firearms and ammunition, they would fire rifles and shotguns to kill or scare off the birds. Today, automated carbide guns and airplanes are used to keep ravenous birds out of the rice fields.

On the *Chênière* Plain, settlers devoted small plots to rice and subsistence agriculture, but ranching was the mainstay of local agriculture. For more than a century, the *chênières* supported a vibrant beef business. Yet Hurricanes Rita and Ike shattered the local industry, decimating livestock holdings on the *chênières* and submerging traditional pastures with saltwater. An important component of Louisiana's agricultural economy was battered, but not destroyed, as ranchers regrouped, bought new heads, monitored their calf crop, repaired their fences, cared for their pastures, drained off saltwater, and went back into the business.

Early mechanization of the Louisiana rice industry, utilizing steam technology, improved the speed and efficiency of moving the harvest from the fields to a mill. (Photo courtesy of the National Archives and Records Administration, Prints & Photographs Division, College Park, MD, photographer Russell Lee, 1938, call number LC-USF33-011639-M2)

Rice was initially marketed in barrels. The introduction of rice sacks permitted producers and millers to reduce production costs and waste. (Photo courtesy of the National Archives and Records Administration, Prints & Photographs Division, College Park, MD, photographer and date unknown, call number 8a28788a)

Table 2 Louisiana sugar mills	
Year	**Number**
1880	1,111
1900	275
1930	70

American markets. Following the Louisiana Purchase (1803), Louisiana's sugar industry had prospered solely because tariffs upon imports—particularly from Caribbean ports—had kept the more expensive and less refined American product commercially competitive. However, after 1898 the new acquisitions were given far greater—and, in the case of Puerto Rico, unfettered—access to American markets.

The resulting flood of imports negatively impacted the Louisiana sugar industry, which fell upon hard times for most of the early twentieth century. Reeling from the glut of imports, the region's sugar farmers watched mosaic disease and other crop illnesses decimate their crops around 1910. According to one observer, this infestation "made such great inroads into the crops of sugar cane during the World War period that Louisiana's sugar industry was threatened with extinction." Between 1911 and 1926, South Louisiana's sugar acreage declined from 310,000 to 129,000 acres—a decrease of 58 percent. The economic repercussions of crop disease and increased competition are seen very clearly in the corresponding reduction of sugar mills in the decades following the Civil War (see Table 2).

These long-standing agronomical and competitive challenges were compounded over the past half-century by new, equally stubborn demographic and economic problems. In the Sugar Bowl parishes, the regional population grows at an annual rate of about 5 percent, and the resulting suburban expansion has increasingly taken long-cultivated fields out of production. In addition, there is intense competition between sugar growers and heavy industry facilities for prime acreage—some of the world's most fertile lands—along the Mississippi River. In the 1960s tax-friendly governmental policies and lax environmental supervision transformed a once pastoral area into "America's Ruhr."

The loss of acreage in traditional production areas has resulted in mill closures, but remaining mills have kept the industry alive by increasing their respective grinding capacities and imposing greater production efficiencies. In addition, the sugar industry has compensated for the loss of prime farmlands by extending cultivation to areas—particularly prairie areas of south-central and southwestern Louisiana—that have traditionally been considered only marginally suited to sugarcane production. As a result, Louisiana's sugar output remains at or near all-time highs.

These robust production figures belie underlying challenges that threaten the industry's continued existence. Mill closures have forced farmers to travel ever greater distances—now often twenty-five to thirty miles or more—to cane-processing sites. The volatility of fuel prices in recent decades threatens the industry's profitability. In addition, regional sugar farmers have lagged behind their counterparts in improving critical planting, harvesting, and crop-handling techniques that have increased growers' profit margins in other sugar-producing areas. Technological stagnation makes farmers more vulnerable to fluctuations in fuel costs and global commodity prices. Finally, the nation's headlong rush into globalization holds the threat of new free trade agreements dismantling the tariffs underpinning domestic sugar industry's continued survival. Aware of these looming threats, some Louisiana sugar growers are hedging their bets by converting productive lands to crawfish ponds or cattle pastures or by growing alternative crops.

The sugar industry's tribulations have been shared by other critical segments of the state's agricultural sector. The cotton industry, slowly rebounding from the devastation of the Civil War, and the emerging rice industry were both staggered by the panic (depression) of 1893 and a perennial glut on the cotton market resulting from foreign competition. Cotton farmers exacerbated the problem by maximizing output to compensate for falling commodity prices, thereby further inundating the already overburdened market and further depressing commodity prices. Furthermore, rice farmers likewise needed to combat the threat of external competition while cultivating regional and national markets.

Louisiana agriculture enjoyed a modest revival during America's involvement in World War I (1917–1918), when commodity prices spiked, and, thanks to greatly increased production in response to the national government's admonitions to increase output, farmers enjoyed profit levels unseen locally in decades. Overproduction for the war effort, however, led to a postwar surplus, and Gulf Coast agriculture stood on the precipice of a protracted crisis. In the mid-1920s a South Louisiana newspaper observed that the region's farmers had not turned a profit since the conclusion of the Great War. The resulting economic distress was compounded by the catastrophic flood of 1927, which literally swept away crops in many areas engaged in row or cover crop production.

Many farmers had not yet repaired the damage wrought by the floodwaters when news of the October 1929 stock market crash presaged the beginning of the Great Depression. Most South Louisiana farming operations would not start to rebound economically until the beginning of World War II (1941). But manpower shortages resulting from the military draft crippled farms of all sizes—particularly in the labor-intensive cotton and sugar industries, which had been slow to mechanize because of their respective persistent economic woes. Wartime prices and the certainty of continued manpower deficiencies for the duration of the war forced the hand of large farming operations.

The resulting mechanization campaign, which began during the war and continued on a progressively greater scale long after the cessation of hostilities in 1945, had an enormous impact upon the regional economic and social landscapes. By the early 1950s machines had displaced thousands of former day laborers and sharecroppers—black and white—who flocked to the nearest towns in search of blue-collar jobs. Local small-scale "freeholders"—unable to make a decent living on their modest agricultural tracts in the postwar era—joined in the exodus, which seems to have peaked after Hurricane Audrey destroyed the coastal plain cotton crop in June 1957. Most of these economic refugees successfully sought well-paying manual labor jobs then emerging in the oil patch. With the departure of the last small farmers in the early 1960s, the region's cotton industry died.

The vacuum was filled by the area's emerging agribusiness industry as venture capitalists—often persons formerly involved in small-to-medium-size plantation operations whose bank accounts were flush with recent oil and gas royalty payments—leased the now dormant small farms and consolidated these properties into massive—though often short-lived—single-crop operations. The failure rate was particularly high among agriculturists experimenting with soybean cultivation after this new row crop was introduced into the area in the late 1960s. By the time the state's oil boom peaked in the early 1980s, the regional soybean industry had largely collapsed. Only the sugar industry survived on the major natural levees and on the prairies, where the crop eventually replaced soybean cultivation to a limited extent, while the rice industry battled for survival in the face of shrinking export markets as a result of Communist insurgency in the Caribbean rim and rising foreign competition.

Along the coast, smaller farming operations reflected in microcosm the experiences of their larger cousins farther inland—but with an important difference. As mentioned earlier, communities of farmers displaced from natural levees consistently turned first to agriculture—their traditional mode of subsistence—as they migrated ever closer to the Gulf shore. Even Cocodrie and Isle de Jean Charles in Terrebonne Parish's coastal marshes were initially farming communities. Indeed, in the early twentieth century Cocodrie supported not only small sugarcane farms but a modest sugar mill as well. Farming in the very heart of the coastal marshes was possible because local freshwater streams sustained both the natural levees and their adjoining prairie or *flotant* buffers with their life-sustaining waters and sediments. The natural cycle of maintenance and renewal, however, was completely disrupted in 1905 as a result of government efforts to mitigate flooding by first damming the Mississippi's distributaries, which, in turn, fed the myriad small waterways that sustained the marsh, and then, after 1927, by channelizing the Mississippi itself. The resulting subsidence and erosion (exacerbated, directly and indirectly, by oil exploration activities—particularly canal dredging—in the marsh), as well as the resultant saltwater intrusion, conspired to bring coastal wetlands agriculture to an untimely end. While monocrop production remains entrenched on the natural levees in the interior, in the coastal communities, it is but a dim and rapidly fading memory shared only by the local elders.

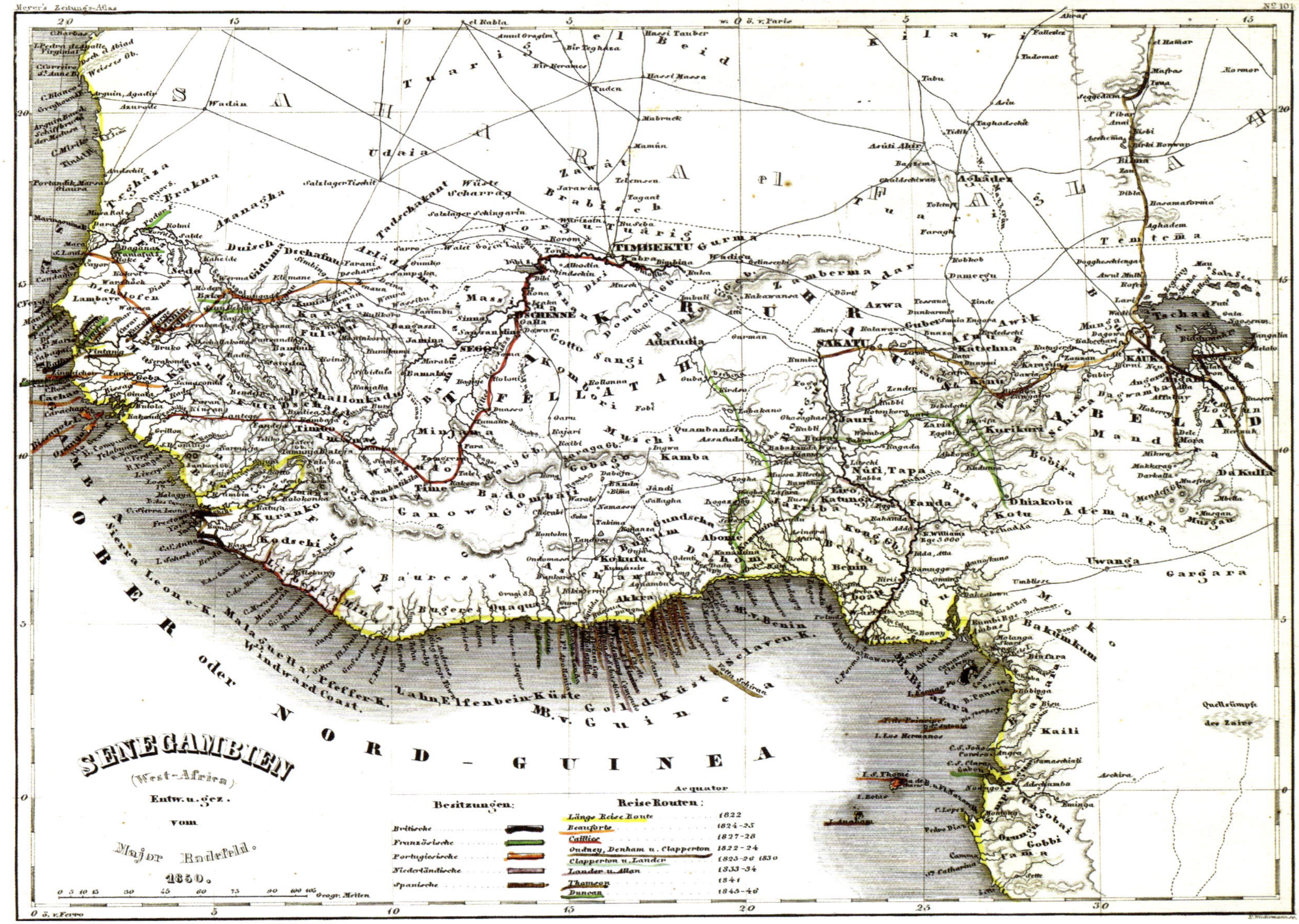

During the French colonial era (1699–1763), most of Louisiana's slaves were drawn from Senegambia, where a subsidiary of the colony's proprietary government (1717–1731) had a monopoly on the commerce in human chattel. (From the authors' collections)

Sugarcane

The "modernization" of the Louisiana sugar industry occurred in lockstep with the American Age of Industrialization. Note the use of barrels and the syrup boat. (Lithograph from *Harper's Weekly*, July 28, 1883, 476)

The emergence of a plantation economy in coastal Louisiana's parishes rapidly polarized the free population economically and politically. By 1859 the yawning divide between the haves and have-nots had precipitated a local civil war between slave-owners and nonslaveholding yeomen that persisted after the Civil War began. Conscripted into the Confederate army against their will, many persecuted nonslave-owners, at the first opportunity, deserted by the thousands. Most found a safe haven in the coastal wetlands. (Lithograph from *Harper's Weekly,* May 14, 1864, 313)

Before modernization, coastal Louisiana's sugar industry was entirely dependent upon manual labor. Large plantations maintained resident laborers and also frequently hired seasonal transient workers, usually small cotton farmers and day laborers who had just completed their harvests. (Photo courtesy of the National Archives and Records Administration, Prints & Photographs Division, College Park, MD, photographer Russell Lee, 1938, call number LC-USF33-011842-M3)

On large plantations, workers loaded cut cane onto narrow gauge "dummy" trains for transportation to mills, while on smaller operations, mule-drawn wagons transported the crop for processing. (Photo courtesy of the National Archives and Records Administration, Prints & Photographs Division, College Park, MD, photographer Russell Lee, 1938, call number LC-USF33-011861-M1)

Sugarcane cultivation was Louisiana's most labor-intensive agricultural industry. Planting, harvesting, and processing required not only backbreaking labor but, in the case of the fall "grinding season," around-the-clock shift work as well. This photo shows female laborers creating windrows in Louisiana's sugarcane fields, ca. 1937. (Photo courtesy of the National Archives and Records Administration, College Park, MD, photographer and date unknown, call number MD, RG 54-Y, Box 1, image Y-242)

This photo records one of the first attempts at mechanization in the coastal Louisiana sugar industry. This machine is burning cut sugarcane stalks before transportation of the crop to the mill. (Photo courtesy of the National Archives and Records Administration, Prints & Photographs Division, College Park, MD, photographer Russell Lee, 1938, call number LC-USF33-011638-M5)

Steam power was first introduced into the Louisiana sugar industry shortly before the Civil War. By the late nineteenth century, mobile derricks were used to load cut cane into barges for transportation to mills. (Photo courtesy of the National Archives and Records Administration, Prints & Photographs Division, College Park, MD, photographer and date unknown, call number 07251pu)

A crane transferring sugarcane from barge to railcars for shipment to a mill. (Photo courtesy of the National Archives and Records Administration, Prints & Photographs Division, Bureau of Plant and Industry, College Park, MD, photographer and date unknown, call number RG 54-Y, Box 1, Y-8421A)

The late 1930s witnessed dramatic changes in the Louisiana sugar industry, as growers began to experiment in earnest with revolutionary new technologies, such as the use of crop dusters. (Photo courtesy of the National Archives and Records Administration, Prints & Photographs Division, Bureau of Plant and Industry, College Park, MD, photographer and date unknown, call number RG 24-B, Box 2, B-211)

In Louisiana, sugarcane is cultivated in three-year cycles. A portion of each crop is reserved as "seed" or "plant" cane, which is cut into stalks (as seen here at Houma in the 1930s) and subsequently placed lengthwise in furrows. Spores located at the cane joints then germinate and sprout. The rapidly maturing plants are then harvested for three consecutive years before the root system is plowed up and replaced. At the end of each three-year cultivation cycle, Louisiana sugarcane farmers are obliged to uproot the cane rhizomes to make way for the planting of new seed cane. In the early twentieth century, this task was accomplished with "stubble diggers." (Photo courtesy of the National Archives and Records Administration, Prints & Photographs Division, Bureau of Plant and Industry, College Park, MD, photographer unknown, 1905, call number RG 54-Y, Box 1, Y-762)

In Louisiana, molasses and syrup have typically been viewed as the lesser sisters of the state's sugar industry. The cane syrup industry was typically maintained for limited syrup production runs for local markets. In the nineteenth century, Deep South residents believed syrup to be an essential antidote to iron deficiency, and contemporary travelers marveled at the amount of syrup these southerners consumed. The Acadia Brand label pictured here targeted South Louisiana's French-speaking population that appreciated the firm's traditional, open-kettle production method. The regional molasses industry, on the other hand, produced far larger quantities for the national market. In fact, during the early twentieth century, one molasses distribution firm maintained a tank farm along the Mississippi near New Orleans for bulk transfers to barges destined for midwestern and northern markets. (From the authors' collections)

Sugarcane mills once dotted the South Louisiana landscape. These "factories" processed sugarcane to produce brown or white sugar, molasses, and syrup for commercial markets. (Photo courtesy of the National Archives and Records Administration, Prints & Photographs Division, Bureau of Plant and Industry, College Park, MD, photographer and date unknown, call number RG 54-Y, Box 1, Y-138)

Built in the early twentieth century to take advantage of New Orleans's port facilities, Domino's Chalmette sugar refinery is the largest sugar refinery in the Western Hemisphere and the second largest in the world. Immediately after Hurricane Katrina, the company's executives created a 200-unit trailer park to house its employees and their families. At a time when most of the region's population was homeless, Domino's employees had a job, family nearby, and a place to live. (Photo by the authors, 2014)

Central Sugar Factory.

The owner of a Bayou Teche sugar plantation, on which there is a central factory, is desirous of leasing the land or forming a co-partnership with some one who has the means to cultivate the same. There are on the place 40 acres of first year's stubble, and seed cane to plant from 150 to 200 acres, or, will lease small tracts for a terms of *years* to those who have the means to improve.

For terms, address X. Y. Z., "Sugar-Bowl" office, New Iberia. July 1-7m

As a result of the economic repercussions of the Civil War and the crippling panic (depression) of 1873, many Louisiana sugar producers were forced out of business. (Advertisement from the New Iberia *Louisiana Sugar Bowl,* September 16, 1888)

During the twentieth century, sugar industry co-ops emerged to sustain the viability of small producers, who cannot survive without mill access. (Photo courtesy of the Center for Louisiana Studies, University of Louisiana at Lafayette, date unknown)

Bagasse consists of residual cellulose fibers remaining after sugarcane stalks are crushed at a mill. The workers pictured here are creating bagasse bales in Houma during the late 1930s. Bagasse was most frequently used as a fuel for both municipal electrical generators and sugar mill boilers and as the principal component of Celotex ceiling tiles. (Photo courtesy of the National Archives and Records Administration, College Park, MD, photographer and date unknown, call number RG 54-Y, Box 1, Y-737)

Laurel Valley Village Plantation in Thibodaux, with more than fifty original buildings in situ, is the largest surviving nineteenth- and twentieth-century sugar plantation complex left in the United States and a visual reminder of the importance of the state's monocrop agriculture. There were three standard plantation layouts developed in Louisiana: the linear, lateral, and block plans. This is an example of a linear plan. The structures are one-room deep, with porches opening to the front, one or two fireplaces for heat, side-facing gables, along with board and batten siding. These design elements were common in the late nineteenth-century slave row houses or cabins. (Photo by the authors, 2004)

Cotton

Before mechanization of the industry in the twentieth century, harvesting cotton was a labor-intensive endeavor. (Photo courtesy of the National Archives and Records Administration, Prints & Photographs Division, College Park, MD, photographer William Henry Jackson, ca. 1885, call number LC-D418-8148)

A truck transporting cotton bales produced on White Lake's south shore. (From the authors' collections, used with permission of the Vermilion Corporation)

When "Cotton Was King," New Orleans was the South's leading cotton exporter, and warehouse space was always at a premium. (Photo courtesy of the Center for Louisiana Studies, University of Louisiana at Lafayette, date unknown)

LOUISIANA.—OUR NATIONAL INDUSTRIES—ARRIVAL OF A CONSIGNMENT OF NINE THOUSAND TWO HUNDRED AND TWENTY-SIX BALES OF COTTON AT NEW ORLEANS.
FROM A PHOTOGRAPH BY E. M. BIDWELL.—SEE PAGE 147.

This lithograph, drawn from a photograph by E. M. Bidwell, depicts a steamboat packed to the gunwales with 9,226 cotton bales. Cotton was a pillar of New Orleans's nineteenth-century economy. (Lithograph from *Frank Leslie's Illustrated Newspaper*, April 30, 1881, page unknown)

Nearly all, or perhaps all, of Louisiana's cotton presses were in New Orleans. For example, the Virginia and Mississippi presses covered two-and-a-half blocks and had the capacity to process 1,000 bales a day. By 1885 the city's register of cotton presses was impressive and dominated the warehouse space that served the port. (Lithograph from *Harper's Weekly*, March 24, 1883, 181)

Stevedores unloading cotton. Cotton was one of the twin pillars of the New Orleans river trade. (Photo courtesy of the National Archives and Records Administration, Prints & Photographs Division, College Park, MD, photographer William Henry Jackson, ca. 1885, call number LC-D418-8121)

By 1873 twenty-six cotton presses existed along Tchoupitoulas Street. The only survivor of this era is a segment of the Amelia Cotton Press building. This section was saved from the wrecking ball, but not before some of the complex was torn town. This section of the building is a visual reminder of cotton's former importance to New Orleans. (Photo by the authors, 2015)

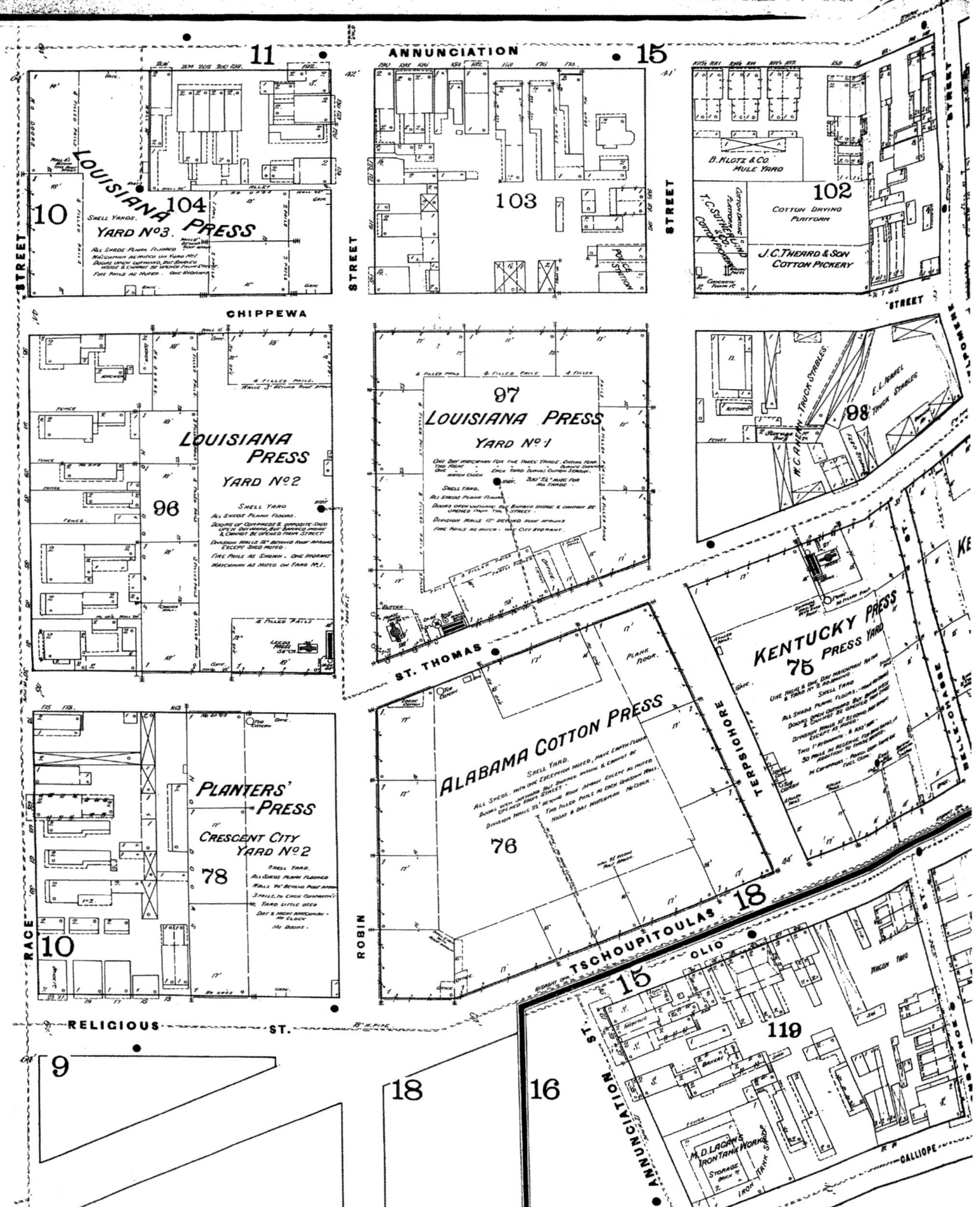

This map indicates the size and number of cotton presses in a small section of late nineteenth-century New Orleans. A thriving drayage industry serviced and supported the press warehouses. A host of drayage operators and their drivers had to endure congested, rutted, narrow, often nearly impassable streets; long lines at the port; and rain delays. (Map from *Sanborn Fire Insurance Maps*, New Orleans, vol. 1, sheet 36 [New York: Sanborn Map and Publishing Company, 1887])

Rice

Rice-Culture in the United States

The First Step—Ploughing

The Second Step—Harvesting

The Last Step—Threshing

A NEW era has dawned for rice-culture in the United States, and it may be said with perfect consistency that in no field of activity are the prospects for future expansion and development more flattering. The new conditions have been produced largely through the instrumentality of the United States Department of Agriculture. The officials of this branch of the national government came to the conclusion some time since that there were latent possibilities in rice-culture, and took steps to foster and restore what was generally regarded at that time as a waning industry.

Dr. S. A. Knapp recently spent eight months in a trip to Japan as an agricultural explorer for the United States government, and made a successful importation of Kiushu rice. The experiments with this rice in the Gulf States during the past season indicate that it is fully twenty-five per cent. more productive than Honduras rice, the variety heretofore grown chiefly in Louisiana, and that its superior milling quality reduces the customary losses by from one-fifth to two-fifths. At a conservative estimate this means a saving to the rice-growers of that State alone of more than $1,500,000 per year. This will give an effective impulse to rice-culture in the United States, particularly to the new system which has lately been developed in southwestern Louisiana.

Another distinct advance is found in the new and improved methods for the irrigation of rice. The discoveries in this direction resulted also from investigations undertaken by the Department of Agriculture in order to overcome difficulties which have up to this time obstructed the progress of the rice industry in this country.

During the past half-century rice production in the United States has grown but little, the crop of 1850, as given by the census of that year, being almost as large as the maximum crop reported since that time, and considerably larger than the average crop of the last ten years. While rice production, as a whole, has remained practically stationary, there has been a decline in the South Atlantic States and an increase in the western Gulf States. Within the past few years the raising of rice in Louisiana and Texas has developed into one of the leading industries of that region, and has given great value to lands heretofore used only for grazing, and to water resources which had been allowed to waste into the Gulf of Mexico.

In this new Eldorado the rice-fields are being handled like the bonanza wheat-farms of Dakota, and fortunes are being made upon comparatively meagre investments. Irrigating canals were started in a small way in Acadia Parish, Louisiana, but development was rapid, and there are now nine separate canal systems in this one parish, one of which, the Crowley Canal, is thirty-five feet wide and eight miles in length, and has ten miles of lateral lines.

Raising Water for Irrigation

Discharging raised Water for the Fields

Elevated Canal for transporting Water to Rice-Fields

The importance of the Louisiana rice industry and its history was outlined in this edition of *Harper's Weekly*. Note the use of mules, bags, wagons, and irrigation canals. (Lithograph from *Harper's Weekly*, November 22 1902, 1742)

The introduction of kerosene- and, later, gasoline-powered tractors permitted an exponential increase in production in Louisiana's rice fields. (Photo from *West Louisiana, East Texas, and the Gulf Coast along the Kansas City Railway*, 1917, Kansas City Southern Railroad, call number Zc22+917we, courtesy of the Beinecke Library, Yale University)

To meet the demands of rice farmers, a number of entrepreneurs developed large pumping facilities and a labyrinth of interconnecting irrigation channels. Canal companies and pumping plants provided water to the farm for a fee, usually a certain portion of the crop. This arrangement continues throughout the prairie's rice farming communities. (From the authors' collections, used with permission of Laura and Buddy Leach)

Rice mills, which stand like skyscrapers on Louisiana's prairies, are constant visual reminders of the cereal grain's regional economic importance. (Photo by the authors, 2015)

Farming

Like sugarcane, citrus trees are highly susceptible to frost and freezing temperatures. Since the late antebellum era, Louisianans residing in the lower coastal plain have supplemented their income through sales of homegrown citrus. The roots of the Pelican State's citrus industry extend to the natural levees below New Orleans, but, by the late nineteenth century, citrus cultivation had extended its geographical footprint to the *Chênière* Plain. A series of unseasonably cold winters in the late 1890s, however, brought an untimely demise to the operations west of Barataria Bay. The industry survives in small agricultural enclaves scattered across St. Bernard, Plaquemines, Lafourche, and Terrebonne Parishes—despite the devastation wrought by successive hurricanes—because of a favorable local microclimate. As a safeguard against freezes, most of the commercial citrus farms are located as close as possible to the Mississippi River or the region's streams and bayous, where they enjoy the advantage of more temperate microclimates associated with these water bodies. Tropical weather systems are more problematic. In recent years, hurricane-force winds and floodwaters have destroyed or severely damaged most of the lower coast's groves. Following a storm, ten years are required to restore replanted orchards to commercial productivity.

Orange Trees for Sale

I have ORANGE TREES from 2 to 5 years old, in a fine healthy condition. They have been tap-rooted and transplanted. I consider this season, from now until February, the best time for replanting orange trees. I can ship these trees to any point, safely and neatly boxed, at reasonable prices. Freight at the cost of buyer. SAMUEL CARY,
Aug19-3m Centreville, La.

Residents of the Louisiana coastal plain cultivated large numbers of citrus trees after the Civil War. (Advertisement from the New Iberia *Louisiana Sugar Bowl*, September 16, 1880)

Because of chronically depressed commodity prices after the panic (depression) of 1873, southern Louisiana farmers experimented with citrus production, and the industry initially flourished, particularly in the coastal parishes. However, unseasonably cold winters in the late 1890s destroyed most of the producing orchards. As a result, across most of the coastal plain, orange production was thereafter confined to household consumption. Louisiana's commercial citrus industry survived only in Plaquemines Parish, where the moderating influence of the surrounding Gulf waters afforded plants greater protection from the cold. (From the authors' collections; photo courtesy of the National Archives and Records Administration, Prints & Photographs Division, Farm Security Administration, College Park, MD, photographer Russell Lee, ca. 1938, call number 8a24461a)

In the late 1870s Plaquemines Parish's most important crops in order of value were sugar, rice, oranges, corn, and farm and garden vegetables for the New Orleans market. In 1878–1879 the parish produced 13,000 hogsheads of sugar, 19,000 barrels of molasses, and 56,000 barrels of rice. At this time, Plaquemines boasted thirty-five active sugar plantations. By the 1930s oranges had become the principal crop, and, over time, the label's advertising, "Louisiana Lower Coast," was identified with Plaquemines Parish. (From the authors' collections)

Perique tobacco comes exclusively from one small region in St. James Parish. At the time the Acadians arrived in the parish in 1766, local Native American tribes were raising this variety of tobacco. Pierre Chenet began raising the crop commercially in 1824. Chenet's nickname was Perique. This photo, from the Poche Perique Tobacco facility, shows how perique tobacco is fermented, as it is pressed under extreme pressure in Convent. (Photo courtesy of PipesMagazine.com)

Although the *Chênière* Plain was formerly noted for cattle and cotton, the region also once produced small quantities of sugarcane. This image depicts a small, family-operated syrup mill at Pecan Island around the beginning of World War II. (Photo courtesy of the Louisiana Department of Wildlife and Fisheries)

Transportation

Before the invention and widespread adoption of corrugated paper boxes, barrels were the containers of choice for Louisiana's shipping industry. In the late nineteenth and early twentieth centuries, millions of barrels moved annually through the Port of New Orleans en route to national and international markets. Barrels were also a staple of commerce in smaller inland ports. For example, in 1888 shippers transported 188,000 barrels of sand and 105,000 barrels of charcoal on the Tchefuncte River. (From the authors' collections)

Louisiana sugarcane farmers face perennial challenges in harvesting their crops. Once cane is exposed to freezing temperatures, the sucrose in the stalks begins to ferment, and growers typically have twenty-four hours to harvest and transport their crop to local mills. Because local roads were notoriously poor, sugarcane farmers along major streams shipped their crops by barge in the early twentieth century. (Photo courtesy of the National Archives and Records Administration, College Park, MD, photographer and date unknown, call number RG 54-Y, Box 1, 928)

Since their introduction into Western commerce, barrels have stored, transported, and aged a wide array of products and liquids. At the height of the barrel trade, stave yards, blacksmiths, and cooperages were required to meet the demand for these sturdy wooden containers. The barrels pictured here were used to sort stretchers used in curing muskrat pelts. (Barrels courtesy of Dot and Mike Benge, Delacroix Corporation; photo by Paula Ouder, Louisiana Sea Grant College Program)

APRIL 14, 1883. HARPER'S WEEKLY. 237

1. Entrance to Atchafalaya River. 2. A "Swamper's" House on the Atchafalaya. 3. A Swamper. 4. Steamer running the Rapids of the Atchafalaya. 5. Red River Landing. 6. Castle on the Atchafalaya. 7. Little Whiskey Bayou. 8. A Swamper's Garden (in a Canoe). 9. The Ash Cabin, Atchafalaya. 10. Map showing Changes in the Mississippi's Current.

A TRIP ON THE ATCHAFALAYA RIVER.—Drawn by J. O. Davidson.—[See Page 235.]

In the nineteenth century, coastal plain cotton growers typically sent their crop to New Orleans brokers (commonly known as factors) by means of steamboats. The journey through Louisiana's bayous was dangerous, expensive, and time consuming because of innumerable navigational hazards. Upon arrival at the Crescent City, cotton bales were offloaded at the "cotton levee," where they awaited sale and transshipment. (From the authors' collections)

The field transportation infrastructure underpinning coastal Louisiana's agricultural industry was sustained by myriad support businesses, including mule dealers, wheelwrights, wagon retailers, trace and harness craftsmen, and farriers. (Photo courtesy of the National Archives and Records Administration, Prints & Photographs Division, College Park, MD, photographer Russell Lee, 1938, call number LC-USF33-011845-M2)

The buggy endured in Louisiana's coastal plain—as in Amish Country—far longer than elsewhere in the United States. In the Bayou Country, its longevity resulted from pervasive poverty. (Photo courtesy of the National Archives and Records Administration, Prints & Photographs Division, College Park, MD, photographer Russell Lee, 1938, call number LC-USF33-011862-M)

Buggies were commonplace on South Louisiana roadways—particularly in the southwestern quadrant—until the late 1940s. (*Rice Belt Journal*, October 26, 1906)

Farmers in the coastal plain generally utilized wagons manufactured in the upper South and the Midwest. (*Rice Belt Journal*, September 28, 1906)

Material Culture

The age of agricultural dominance witnessed the emergence of several distinctive indigenous architectural styles, including Acadian/Cajun (upper left), Creole (upper right), and shotgun (lower left). Early forms of each style were consistently internally asymmetrical, because pioneer Louisianans considered foyers and halls to be lost space. Louisiana's antebellum era (1815–1860) also witnessed the introduction and flowering of Greek Revival architecture (lower right). The Greek Revival structure pictured here, Evergreen Plantation House, was resurrected from near ruin by Matilda Geddings Gray. The plantation complex boasts thirty-seven historic structures. (Upper right photo courtesy of the Center for Louisiana Studies, University of Louisiana at Lafayette; all other photos by the authors, 1978, 2009, 1989)

Although described as the French Quarter—or *Vieux Carré*—the seventy-eight-square-block neighborhood's architectural roots date from the Spanish colonial period. The Great New Orleans Fires of 1788 and 1794 destroyed most of the Quarter's standing French colonial structures. The Spanish rebuilt the city in compliance with strict building and fire codes. Buildings consequently exhibit Iberian building traditions, including elaborate ironwork on many second-floor balconies. (Photo by the authors, 2013)

A raised Creole cottage was designed to survive floods by using the first floor for storage and the second as living quarters. Note the shaded porch, three chimneys, cistern, and pyramidal roof with one small dormer. (Photo courtesy of the National Archives and Records Administration, Prints & Photographs Division, College Park, MD, photographer Russell Lee, 1938, call number LC-USF33-01615-M4)

New Orleans boasts a vast array of house types. One of the most distinctive is the raised basement house, which was designed to withstand severe flooding. (Photo by the authors, 2007)

Rainwater collected in stand-alone cypress cisterns was an essential source of potable water in the nineteenth and twentieth centuries. (Photo courtesy of the National Archives and Records Administration, Prints & Photographs Division, College Park, MD, photographer and date unknown, call number 8a23582a)

Mud ovens were commonplace in rural southeastern Louisiana's wetland region, where wheat, imported from New Orleans, was relatively cheap. West of the Atchafalaya River, freight charges made wheat flour a luxury. (Photo courtesy of the National Archives and Records Administration, Prints & Photographs Division, College Park, MD, photographer Ben Shahn, 1935, call number LC-USF33-006174-M2)

A South Louisiana farm complex framed by a *pieux* fence. (Photo courtesy of the National Archives and Records Administration, Prints & Photographs Division, College Park, MD, photographer Russell Lee, 1938, Farm Security Administration, call number 8a24204a)

In the first half of the twentieth century, amenities like school boats, pictured here, permitted coastal dwellers to reside in areas for exploitation of wetland resources and still allow young people to obtain an education. (Photo from the Colonel Joseph S. Tate Photograph Album, Mss. 4963, Louisiana and Lower Mississippi Valley Collections, LSU Libraries, Baton Rouge, photo number 4963024; hereafter cited as Tate Album, Mss. 4963, LLMVC)

Before batteries, small kerosene headlamps were used by alligator hunters to immobilize their prey. (Photo by Paula Ouder, Louisiana Sea Grant College Program)

Our Lady of LaVang Catholic Church is a Vietnamese icon in South Louisiana. (Photo by the authors, 2013)

Armies of axmen had to work long hours to feed the insatiable demand for cypress logs at regional mills. (From the authors' collections, used with permission of Rathborne Properties)

3

INDUSTRIALIZATION

The protracted decline of coastal Louisiana agriculture—both monocrop (large-scale commercial) and diversified (small-scale yeoman) farming—coincided with the rise of Gulf Coast industrialization. In fact, the former precipitated the latter as locals consciously sought alternative means of gainful employment.

This is not to say that there was no industrial activity in Louisiana prior to agriculture's waning fortunes. Far from it. In fact, between 1899 and 1909 the number of industrial facilities in the Pelican State increased from 1,826 to 2,516 (a 37 percent rise). In the latter year, Louisiana ranked nineteenth nationally in terms of the gross value of its manufactured products. In addition, the US Department of Commerce reported that "from 1849 to 1909 the value of [Louisiana's] manufactures [had] increased at a somewhat greater rate than the value of the manufactures of the United States as a whole."

Coastal plain residents first engaged in industrial activities—albeit on a modest scale—during the colonial era. Industrialization initially centered upon refining natural resources, processing agricultural products and by-products, and harvesting, processing, and distributing the coastal wetlands' renewable natural resources. After a failed experiment with silk production, involving imported silkworms and native mulberry trees in the 1720s, colonists on the natural levees turned their attention to commercial indigo and tobacco. In the prairies, where existing implements could not penetrate the shallow claypan, settlers turned to ranching, which, in turn, created a regional beef industry. In the late 1770s and early 1780s Spanish Louisiana mounted a sustained invasion of British West Florida in support of the Patriot cause during the American Revolution. The colonial army was sustained in the field by beef driven overland from the Attakapas and Opelousas areas to processing sites along the Mississippi.

In the late eighteenth and early nineteenth centuries, prairie cattle became the principal source of beef for the Crescent City. During the antebellum steamboat era, Washington (St. Landry Parish) and New Iberia (Iberia Parish) emerged as primary preprocessing and shipping points. Beef slaughtered to meet local demand provided hides to large local tanneries and, in St. Landry Parish, saddle manufacturers. Most of the cattle, however, were loaded on barges for delivery to the massive slaughterhouses

Before the emergence of steam-powered sawmills, axmen used a two-man crosscut saw along with a froe, mallet, and perhaps a small bucksaw to produce shingles. (Lithograph from *Picturesque America* [1872], nonpaginated)

along the Mississippi downriver from New Orleans's French Quarter. These shipments, eventually supplemented by herds from Texas, sustained the municipality's emerging meatpacking industry and budding local shoe and boot factories.

The coastal plain's thriving cattle commerce ended with the onset of the Civil War and the disruption of intrastate trade following the Union occupation of New Orleans in the spring of 1862. In the wake of the Civil War, the occupation and fencing of the open range in Louisiana's prairie parishes, along with the establishment of railroad communications between the Great Plains and Chicago, which diverted cattle shipments to the upper Midwest, particularly the Windy City, brought a speedy demise to the prairie cattle business. In ensuing decades, the regional ranching industry would retreat to the coastal marshes, where luxuriant open grasslands were readily available.

In other areas not conducive to husbandry, coastal citizens focused their attention on woodland resources. Residents of Lake Pontchartrain's north shore, according to a report by Governor Jean-Baptiste Le Moyne de Bienville, had established a naval stores industry by 1734. Operating out of three or four "plants," workers gathered pine sap from which were derived modest quantities of turpentine, pitch, and tar for regional consumption and export. Louisiana's naval stores industry survived into the nineteenth century, when it was sustained by an explosive growth in demand for its products in the neighboring New Orleans market.

Yet the regional pine sap–based products industry remained small in scale, with the exception of a brief period at the turn of the twentieth century, when it enjoyed a short-lived growth spurt. Between 1899 and 1909, the region's turpentine production increased from 219,504 to 1,231,254 gallons, while rosin output rose from 23,843 to 139,480 barrels. However, the uptick in production resulted in the depletion of local pine forests, which was compounded by destruction of "much of the timber which was being worked" during the 1909 hurricane. These devastating developments resulted in greatly curtailed subsequent production.

Settlers along the Mississippi River, on the other hand, began to harvest wood—primarily cypress—to satisfy both the domestic demand for construction materials

Early Louisiana surveyors remained vigilant for oak timber suitable for construction of naval warships; however, scholars have yet to determine the boundaries of Louisiana's federal live oak naval reserves. (Lithograph from *Harper's Weekly,* January 19, 1884, 44)

In the early eighteenth century, French colonial officials scrupulously reported the availability of oak—particularly live oak—timber, a precious resource in constant demand for naval warship construction in the metropole. Although lower Louisiana had an abundance of this commodity, labor and environmental issues prevented the commercial exploitation of this product—at least the legal exploitation of this resource—for the remainder of the wooden ship era. "Because it lacked a deep-water port, the navy never considered Louisiana a prime site for maritime construction. The state did possess, however, an abundant supply of valuable live oak which could be utilized in Atlantic coast shipyards. After making surveys in 1815 and 1819, the Navy placed 19,000 acres of live oak on reserve in Louisiana. Yet little timber was cut by the government, because America's acquisition of Florida from Spain in 1819 provided the East Coast shipyards a more convenient source. Hence, the government made feeble efforts at best to guard Louisiana's federal live oak reserves, and, as a result, thieves stole millions of cubic feet annually—usually for sale to the United States Navy!"

Source: W. G. Piston, "Maritime Shipbuilding and Related Activities in Louisiana, 1542–1986," *Louisiana History* (1988): 163–175.

Logging brought the first wave of industrialization to Louisiana's coastal wetlands. (From the authors' collections, used with permission of Rathborne Properties)

and the export market. Most of the lumber shipped abroad usually was consumed by Saint-Domingue (present-day Haiti), Louisiana's principal eighteenth-century trading partner. After the catastrophic fire of 1785, Saint-Domingue colonists used Louisiana timber to reconstruct the colonial capital, Cap Français, while their fellow Louisianians used locally harvested and processed cypress to rebuild New Orleans following the destructive conflagrations of 1788 and 1794 and the devastating hurricanes of 1787, 1793, 1794, and 1800. In Gulf Coast and Caribbean locales, old-growth cypress lumber was the near-perfect building material, for it was resistant to perpetually high humidity, ground moisture, and native termites; it was also somewhat less flammable than other wood varieties.

Colonial Louisianians initially dressed cypress timber either with crude pit saws or by means of mechanical saws powered by flumes fed by the Mississippi River's seasonal floodwaters. Processing raw timber by either method became more difficult over time as lumbermen exhausted the most accessible cypress stands in the swampy waterbottoms of their land grants. To meet the demand, axmen began to seek out new sources of supply ever deeper in the forbidding and hazardous wetlands. Because of the steadily growing distance between the cypress groves and the processing areas on the natural levees, lumbermen devised an ingenious method of transporting the enormous old-growth logs to their primitive "mills."

Moving into the swamps during the "high-water season" (typically January–June), when the Mississippi's floodwaters inundated the coastal lowlands, colonists girdled harvestable trees, removing a band of bark around the lower tree's circumference, in the process killing the plant, draining its sap, and making the moribund log buoyant. The loggers returned in the dry, cool fall months, when the swamp bottoms

Louisiana's cypress mills typically produced raw and finished products, including shingles and railroad ties. (From the authors' collections, used with permission of Rathborne Properties)

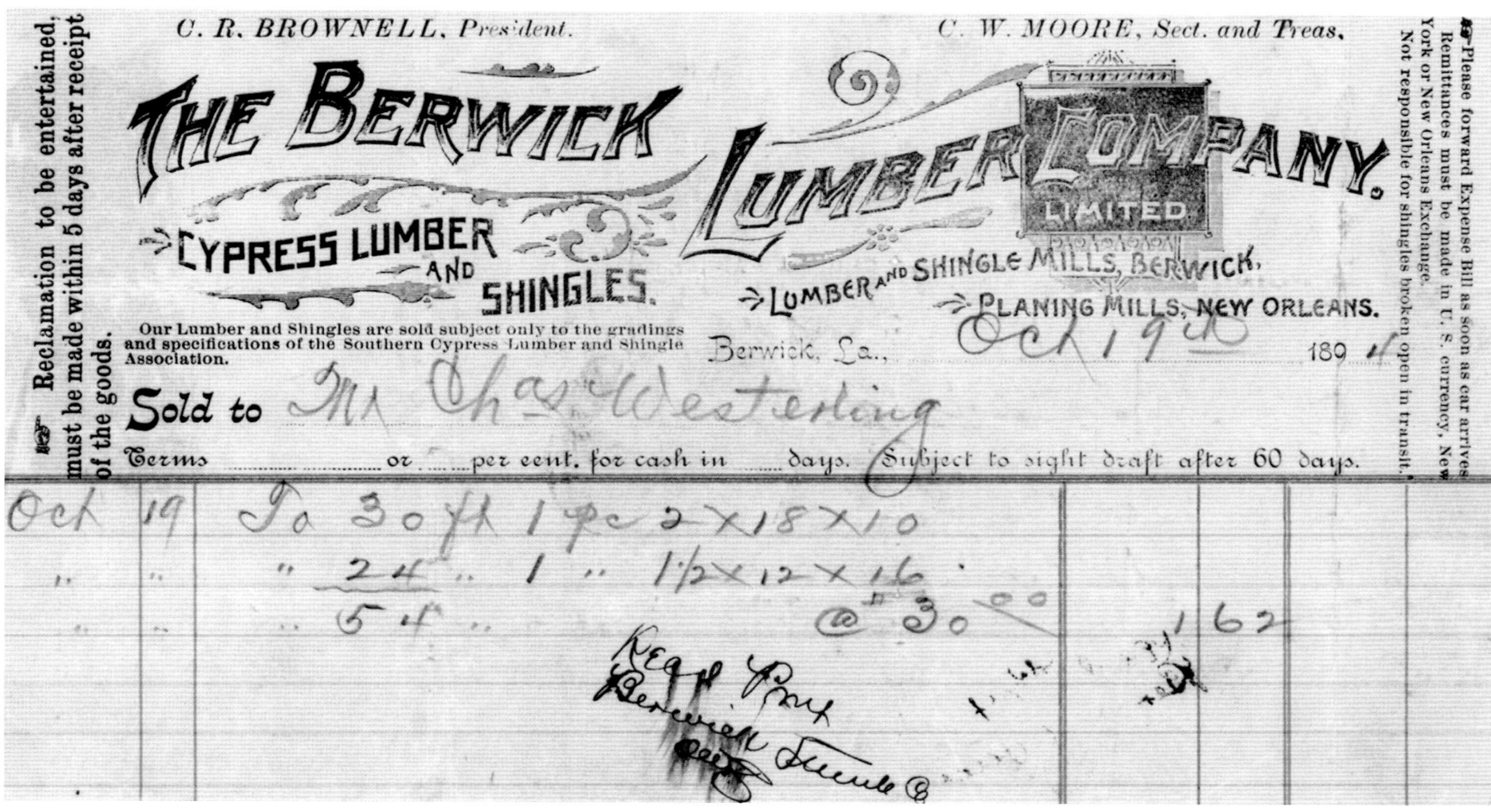

Reclamation to be entertained, must be made within 5 days after receipt of the goods.

C. R. BROWNELL, President. C. W. MOORE, Sect. and Treas.

THE BERWICK LUMBER COMPANY. LIMITED

CYPRESS LUMBER AND SHINGLES.

LUMBER AND SHINGLE MILLS, BERWICK. PLANING MILLS, NEW ORLEANS.

Our Lumber and Shingles are sold subject only to the gradings and specifications of the Southern Cypress Lumber and Shingle Association.

Berwick, La., Oct 19 1894

Sold to Mr Chas. Westerling

Terms .... or ...... per cent. for cash in ...... days. Subject to sight draft after 60 days.

Please forward Expense Bill as soon as car arrives. Remittances must be made in U. S. currency, New York or New Orleans Exchange. Not responsible for shingles broken open in transit.

It was not uncommon for a cypress lumber company to operate mills close to both wetland harvesting sites and the region's largest urban market. Further, the shingle trade—now largely forgotten—was once an important integral part of the Louisiana cypress lumber industry. In late June 1898 the *Lafayette Advertiser* reported that Louisiana mills had a *daily* output of 2,475,000 shingles. (Courtesy of the Morgan City Archives, Morgan City, LA, 1894)

were relatively dry, to cut and limb the girdled trees. Loggers allowed the harvested saw-timber to remain on the ground until the next vernal flood, when they moved the floated saw logs to landings on the natural levees, where ox teams could haul them to rustic mills.

This system remained the basis for the cypress industry's field operations for the remainder of its existence, but the introduction of vastly improved harvesting, transportation, and milling technologies after the Louisiana Purchase (1803) permitted cypress commerce to grow exponentially, eventually dominating the regional economic landscape.

The industry really came into its own in the post–Civil War era, when the nation's major logging companies, hard on the heels of clear-cutting the coniferous forests of the upper Midwest, relocated to Louisiana's remaining old-growth cypress groves. The virgin, commercial-grade timber within these even-aged stands was located

A number of cypress companies owned their own fleet of steamboats to move logs from wetland harvesting sites to mills. (Photo courtesy of the Center for Louisiana Studies, University of Louisiana at Lafayette, date unknown)

Loggers used chain-dogs, like the one pictured here, to bind floating cypress logs into rafts. (From the authors' collections, used with permission of Charles and Linda Landry)

primarily in the remote reaches of the Atchafalaya Basin and in the Chacahoula (near Thibodaux) and Manchac (near New Orleans) swamps. These nascent large-scale felling and milling operations quickly supplanted the far more modest established homegrown ventures. Benefiting from the national home construction and railroad industries' "discovery" of "the wood that defies decay," the cypress lumber industry experienced phenomenal growth between 1899 and 1909—an estimated 218.5 percent in lumber production and 50.1 percent in shingle output.

A paradigm shift in field harvesting techniques resulting from the convergence of new technologies permitted such explosive growth. In the 1890s William Baptist of New Orleans invented a pullboat logging skidder that "snaked" massive cut logs to a central collection point. This floating platform featured powerful steam winches capable of pulling felled logs to gathering and assembling sites from a distance of up to 3,000 feet along laser-straight pathways.

After pullboats were transported to strategic locations in the heart of Louisiana's most remote swamps via a network of freshly dug timber access canals (some of which were seven miles long, with numerous feeder branches), logging crews established "drag paths" often radiating nearly 360 degrees from the winch site like spokes from a wheel's hub. (More than a century later, the landscape still bears the scars of these logging trails or runs.) These logging techniques permitted the near-complete removal of old-growth cypress.

Logs assembled and limbed at the center of the wheel's hub were then transported to mills by one of two methods: In areas where logging canals afforded easy access to natural waterways, loggers organized buoyant cypress trunks into large rafts held

In its heyday (1881–ca. 1925), the F. B. Williams Mill at Patterson was reputedly the world's largest cypress sawmill. (Masthead courtesy of the Morgan City Archives, Morgan City, LA, 1900)

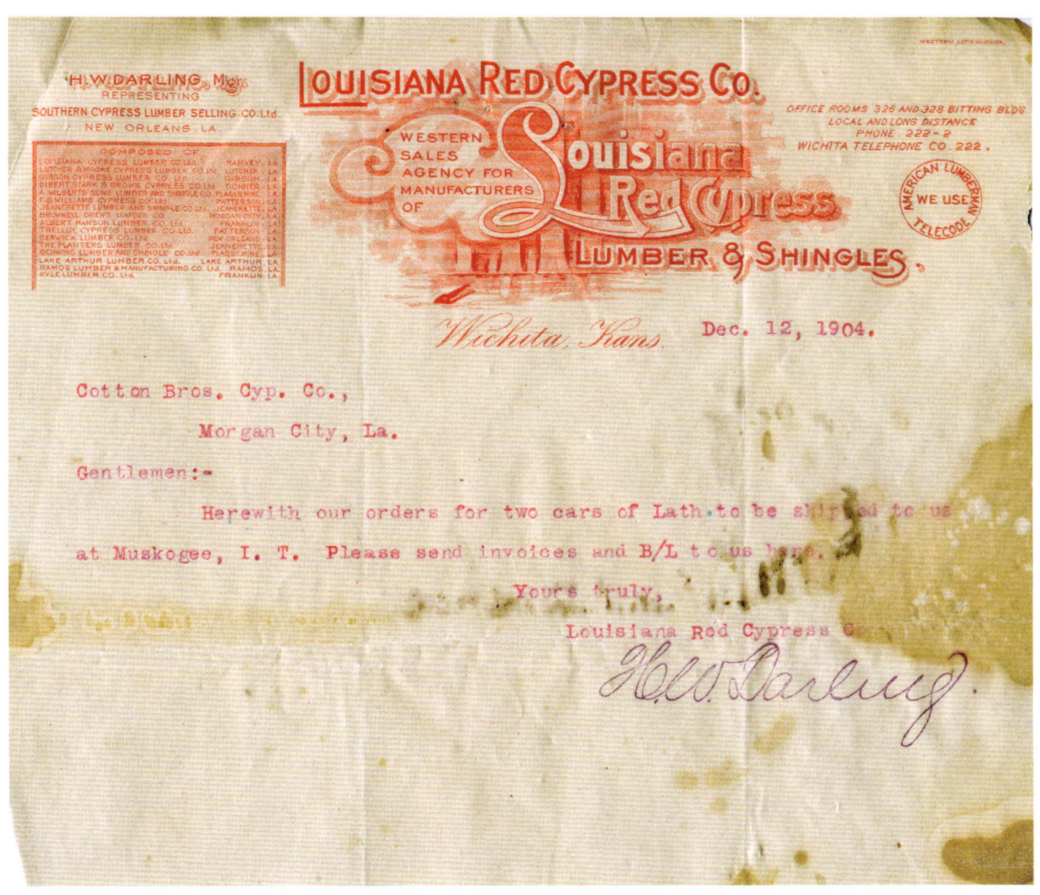
H. W. DARLING, Mgr.
REPRESENTING
SOUTHERN CYPRESS LUMBER SELLING CO. Ltd
NEW ORLEANS, LA.

COMPOSED OF
LOUISIANA CYPRESS LUMBER CO. Ltd. HARVEY, LA.
LUTCHER & MOORE CYPRESS LUMBER CO. Ltd. LUTCHER, LA.
GIBSON CYPRESS LUMBER CO. Ltd. GIBSON, LA.
DIBERT, STARK & BROWN CYPRESS CO. Ltd. DONNER, LA.
A. WILBERTS SONS LUMBER AND SHINGLE CO. PLAQUEMINE, LA.
F. B. WILLIAMS CYPRESS CO. Ltd. PATTERSON, LA.
JEANERETTE LUMBER AND SHINGLE CO. Ltd. JEANERETTE, LA.
BROWNELL DREWS LUMBER CO. MORGAN CITY, LA.
ALBERT HANSON LUMBER CO. Ltd. FRANKLIN, LA.
TRELLUE CYPRESS LUMBER CO. Ltd. PATTERSON, LA.
BERWICK LUMBER CO. Ltd. NEW ORLEANS, LA.
THE PLANTERS LUMBER CO. Ltd. JEANERETTE, LA.
SCHWING LUMBER AND SHINGLE CO. Ltd. PLAQUEMINE, LA.
LAKE ARTHUR LUMBER CO. Ltd. LAKE ARTHUR, LA.
RAMOS LUMBER & MANUFACTURING CO. Ltd. RAMOS, LA.
KYLE LUMBER CO. Ltd. FRANKLIN, LA.

LOUISIANA RED CYPRESS CO.
WESTERN SALES AGENCY FOR MANUFACTURERS OF
Louisiana Red Cypress
LUMBER & SHINGLES.

OFFICE ROOMS 326 AND 328 BITTING BLD'G.
LOCAL AND LONG DISTANCE
PHONE 222-2
WICHITA TELEPHONE CO. 222.

AMERICAN LUMBERMAN TELECODE
WE USE

Wichita, Kans. Dec. 12, 1904.

Cotton Bros. Cyp. Co.,
Morgan City, La.

Gentlemen:-

Herewith our orders for two cars of Lath to be shipped to us at Muskogee, I. T. Please send invoices and B/L to us here.

Yours truly,
Louisiana Red Cypress Co.
H. W. Darling.

The masthead of the Louisiana Red Cypress Company's Wichita, Kansas, branch office. (Masthead courtesy Morgan City Archives, Morgan City, LA, 1904)

together by "devil dogs"—chains with spiked ends; specially adapted steamboats then towed the rafts to sawmills intentionally located on the peripheries of old-growth stands. When waterborne transportation was not feasible, logging syndicates constructed temporary narrow-gauge "floating" railways into sites with limited water access. Located primarily east of the Mississippi River in the Lake Maurepas Basin, these railroads featured rails mounted on floating log mats. (Contemporaries often referred to this extractive process as "liquid logging.") Once the logging operation concluded, laborers dismantled the tracks, leaving behind narrow passageways the clear-cut forests eventually reclaimed. Historical quadrangle maps and rare photographs now provide the only evidence of their existence.

The steady stream of timber removed from ever more remote recesses of the coastal plain's dreary, hostile, and forbidding waterbottoms supported enormous

Completion of a railway across south-central and southwestern Louisiana in 1880 inadvertently gave birth to a niche industry that has, over time, emerged as an important regional economic driver. Prior to the coming of the Iron Horse, St. Martinville had been the terminus of a public roadway established earlier in the nineteenth century to facilitate the movement of agricultural produce from Lafayette and upper St. Martin Parishes to the head of navigation on Bayou Teche. Bypassed by the critical new communications artery, however, St. Martinville was forced to cast about for some means to revive its now seemingly doomed economic fortunes. The city fathers settled upon a solution that was revolutionary for the time—cultural tourism. Henry Wadsworth Longfellow's epic *Evangeline*, an unprecedented literary phenomenon, made the story of the Acadian exile an integral part of America's literary canon. The Evangeline story line was significantly altered in *Acadian Reminiscences*, a mid-1890s novella by Felix Voorhies. Published first in serialized form in local newspapers and later in book form in 1907, the Voorhies version recast the ending of the Longfellow's epic, setting the reunion of Evangeline and Gabriel in St. Martinville. Not surprisingly, the story was an immediate literary sensation in Acadian/Cajun South Louisiana. St. Martinville capitalized upon the resulting excitement by establishing an Evangeline park featuring an oak under which the fictional reunion occurred, and began billing itself as South Louisiana's premier tourist destination site. Tourist excursions caused an immediate upturn in the local economy. The community continued to build upon this success by lobbying for the establishment of a Longfellow-Evangeline national park at St. Martinville. Although this effort fell short of its goal because of the onset of the Great Depression in 1929, it did result in the establishment of the first state park (now known as the Longfellow-Evangeline State Historic Site), which constituted the centerpiece of the regional rural tourism industry for two generations.

The development of a tourism industry in St. Martinville paralleled in microcosm the development of a tourism economy in New Orleans. During the early-to-mid-nineteenth century, the Crescent City's cultural attractions—consisting primarily of theaters and the famous French Opera House—as well as racetracks and top-tier hotels and restaurants attracted well-heeled antebellum travelers. Following the Civil War, male business travelers sought out very different attractions—in the Crescent City's legalized prostitution district of Storyville. The city closed Storyville at the behest of the US Navy in 1917, and, in an effort to rehabilitate its now tarnished national reputation as a den of iniquity, it began to promote itself as a showcase for the "quaint" architecture of the French Quarter, which, in the early post–World War I era, had become a glorified slum populated by bohemian artists and writers, including William Faulkner. In 1925 the New Orleans city council established the first Vieux Carré Commission to safeguard preservation of the Quarter's unique architectural heritage.

steam-powered sawmills. These mills manufactured staggering quantities of shingles, barrel staves (also called shook), railroad crossties, and lumber for national distribution. In 1909 Louisiana "was the second state in the Union . . . in the amount of lumber sawed." That year, the Bayou State, which boasted "some of the largest sawmills in the country," shipped 608,854,000 board feet (boards that measured twelve inches by twelve inches by one inch thick) of cypress lumber. In addition, by 1916 Louisiana cypress crossties also underlay much of the 254,000 miles of railroad tracks crisscrossing the United States. Still more lumber was diverted to the coastal plain's shipyards—from the massive industrial installations in the New Orleans area to the folk shipyards along coastal bayous—and the emerging commercial fishing industry's infrastructure (warehouses, docks, wharfs, and drying platforms). However, Louisiana's finite cypress resources could not sustain such unbridled output, and the region's once vast old-growth stands were exhausted by the beginning of World War II.

Other exploitative Louisiana industries experienced identical evolutionary trajectories, enjoying both a meteoric rise as a result of technological innovations and an equally spectacular, devastating collapse as local resources were systematically harvested until depletion. Louisiana's Spanish moss industry, for example, was directly tied to the cypress industry's fortunes. As a result of notable regional environmental changes, moss harvesting and processing first rose to economic prominence in the nineteenth century. Significant sustained changes in the Atchafalaya Basin's hydrology during the postbellum era—resulting in large part from wartime damage to the region's rudimentary levee system and the consequential inundations—forced many of the area's subsistence farmers to find other livelihoods. Many eventually settled upon seasonal folk occupations, including fishing, trapping, and Spanish moss gathering.

Spanish moss is an epiphyte—a nonparasitic plant growing on other flora and drawing its nutrients from the air—related to the pineapple. The plant consists of an organic exterior and an inorganic fibrous interior. The species once thrived in the Louisiana coastal plain, particularly in the area's cypress groves, and it was in constant demand by a multitude of late nineteenth- and early twentieth-century industries.

Market hunters were the terrestrial counterparts of commercial fishermen. Before the age of refrigeration, since ice was quite expensive in rural areas, commercial hunters field-dressed game birds before stuffing the carcasses with Spanish moss and packing them in barrels—approximately 100 per container—for shipment to rural rail stops. (Hunters created alternating rows of birds and salt in the barrels.) At the depots, workers iced down the barrels and then forwarded them to New Orleans, where French Market vendors displayed them for sale to the Crescent City's highly discriminating consumers. Extant documentation indicates that, in a typical year, New Orleanians consumed nine to ten commercially harvested birds per capita. This system endured for decades until passage of the 1900 Lacy Act and ratification of the 1918 Migratory Bird Treaty ended the wanton slaughter of species driven to the verge of extinction.

More than 325 bird species migrate annually along the Mississippi Flyway, and, as a consequence, Louisiana, the flyway's southern terminus, is popularly known as a "Sportsman's Paradise." The superabundance of game birds gave rise early on to a deeply entrenched hunting culture in the Louisiana coastal plain. Successful duck hunting required the use of decoys, which, in turn, engendered one of the first truly American art forms. The pictured decoy is a representative example of the award-winning work by the Bayou State's highly skilled woodcarvers. (Carving courtesy of Dot and Mike Benge, Delacroix Corporation; photo by Paula Ouder, Louisiana Sea Grant College Program)

Rafts of cypress logs waiting processing at an unidentified waterfront lumber mill. (From the authors' collections)

Waterways, whether natural or manmade, had to be constantly maintained by dredging. (Photo from Tate Album, Mss. 4963, LLMVC, photo number 4963047)

First used as mattress stuffing by pioneers within the coastal plain, Spanish moss later came to be valued for its uses as packing material, floral filler, mulch, insulation, and stuffing for premium-quality furniture. The demand for cured and processed moss increased steadily after the Civil War as cypress logging operations opened vast moss-covered cypress stands to industrial harvesting.

Swampers, who viewed moss as a "lagniappe crop" and welcomed seasonal employment in the fall and winter months, gathered moss by means of hook- or barb-tipped poles from standing skiffs outfitted with a *jourg* (sometimes rendered *joug*). In some swamps, cypress logging companies allowed moss-gatherers to ride their "floating railroads" to prime logging sites in exchange for a fee or a percentage of the moss removed from felled trees.

Moss gatherers transported the green moss to processing yards, where the epiphyte was heaped into conical piles to initiate the curing process. To promote decay and prevent spontaneous combustion, harvesters periodically turned—or "worked"—the moss, until the pile turned a deep brown. Laborers then transferred the residual, tough moss fibers to fences for drying. When the brown fibers became black, the curing process was complete. Workers subsequently removed twigs, leaves, and other detritus from the moss and gathered the fibers for ginning. Ginning facilities graded, laundered, and baled the moss for export to manufacturers, usually in urban centers.

Coastal Louisiana's Spanish moss industry—like the cypress industry to which it was intimately tied—peaked in the early twentieth century. By 1930 there were at least eighteen gins in lower Louisiana, and in 1927 alone the area shipped approximately

1,200 railroad carloads of ginned moss to manufacturers. However, output declined sharply with the demise of the regional cypress industry in the 1930s and early 1940s. Disappearance of cypress groves—the principal source of commercial-grade moss fiber—was compounded by the virtual disappearance of Spanish moss in much of South Louisiana as a result of post–World War II–era air pollution. By the late 1970s only one moss gin remained, and a decade later the industry was extinct. (The manufacturers of foam rubber opportunistically filled the void.)

Like moss gathering, commercial hunting also began as a relatively modest undertaking in the colonial era as enterprising colonists killed and field-dressed wild game for sale in New Orleans markets. Because early market hunters were dependent upon pirogues and other paddle-powered craft and due to the highly perishable nature of the commodity, field operations were necessarily limited geographically to a small radius around the Crescent City's French Quarter. However, with the completion of railways into St. Bernard and St. Mary Parishes in the 1850s and into southwestern Louisiana in the 1880s, the commercial hunting industry's reach expanded exponentially. Thanks to the ready availability of wildfowl in local markets and restaurants, New Orleans's population had clearly developed a taste for game birds, but as a result of the city's amazing demographic growth in the mid-nineteenth century, local hunters could no longer meet the rapidly growing demand. By the mid-1880s commercial hunting operations targeting New Orleans consumers began to emerge in wetland areas newly serviced by railroads—primarily the south-central and southwestern quadrants of the coastal plain. These fledgling operations, evidently working

The New Orleans French Market was the principal destination of most of the game birds killed by commercial hunters as well as much of the seafood harvested by regional fishermen. (Photo courtesy of the National Archives and Records Administration, Prints & Photographs Division, College Park, MD, photographer and date unknown, call number unknown)

About the size of a rice irrigation channel or a drainage ditch, the *trainasse,* or pirogue trail, provided trappers access to their trapping leases. (Photo courtesy of the Louisiana Department of Wildlife and Fisheries)

in cooperation with Crescent City business interests, shipped prodigious numbers of game birds to the city by rail.

For example, on September 4, 1886, the *Rayne Signal* reported that "half a million wild ducks are annual[ly] killed in Southern Louisiana and sent to the New Orleans market." And the wanton slaughter would continue unabated for decades. In fact, it would actually increase by the turn of the twentieth century. During the 1908–1909 hunting season, Calcasieu Parish's commercial hunters alone bagged over *one million* game animals, including 600,000 ducks, 300,000 snipe, 35,000 quail, and 25,000 geese—valued at about $170,000 (approximately $4,524,192 in 2015 dollars).

The wanton destruction of wild game was not limited to fowl consumed for their meat. During the late nineteenth and early twentieth centuries, feathers were an essential component of fashionable dress. They were, in the words of one commentator, "the height of fashion." Trendy Victorian hats featured not only a broad array of feathers but stuffed bird carcasses as well. The fashionistas' feather fetish reached its height during the first decade of the twentieth century, when more than fourteen million pounds of plumage reached London fashion houses, while their counterparts in Paris and New York received consignments of comparable scale.

In New York, the industry enjoyed phenomenal growth during the first decade of the twentieth century. The milliners' success, in turn, engendered a 357 percent rise in the value of processed plumage products between 1904 and 1909. In the latter year, New York, which accounted for 88.2 percent of the plumage manufacturing in the nation, boasted 319 "feather and plumes" establishments (77 percent of the national total) with 9,813 employees and an annual output valued at $23,980,567 ($612,642,197 in 2014 currency).

The uncontrolled slaughter underpinning feather consumption of such unprecedented magnitude resulted in the extinction or near extinction of numerous avian species. including, respectively, the passenger pigeon and egret. This carnage led directly to the adoption of state and federal conservation laws that brought a swift, merciful end to both the plumage trade and commercial hunting. Again, seemingly inexhaustible resources had been extinguished in little over a generation.

In 1889 the US Census Bureau identified Florida, Louisiana, and Texas as the Gulf Coast states most extensively involved in the plumage trade. The Pelican State ranked second in this troika, exporting 78,328 birds—22.66 percent of the regional total. Over 58,000 of these wildfowl were of unidentified varieties, but white (12,220) and sandhill (7,017) cranes constituted more than 95 percent of the documented species.

The systematic exploitation of other fauna fortunately proved to be more sustainable—largely the result of governmental conservation regulations and consistently vigorous enforcement efforts that curbed overharvesting. This is particularly true of coastal Louisiana's once robust trapping industry. The Acadian ancestors of the region's dominant Cajun population engaged in seasonal trapping from the time of their arrival in Canada's Bay of Fundy basin in the early seventeenth century. The Acadians continued this tradition following their expulsion from Canada in 1755 and their subsequent migration to the lower Mississippi Valley. However, the fur-bearing animals that they encountered in their adopted semitropical homeland were far less valuable and marketable than the Canadian species they formerly stalked.

As a result, fur trapping appears to have continued in the coastal plain, but only for home consumption. In fact, in its analysis of the 1880 census, the US Census Bureau did not even list Louisiana among the states producing dressed furs. While there was modest trapping activity focusing on muskrat and otter pelts in the early twentieth century, the Louisiana fur trapping industry would not truly blossom until the Roaring Twenties, when a new fashion craze—men's ankle-length fur coats—made Louisiana's muskrat pelts stylish.

Between 1914 and 1922, market prices for muskrat pelts rose from $0.08 to $0.50 per pelt, and trapping became more profitable than hunting for the first time. A single fur was twice as valuable as a brace of ducks in the early 1920s. Pelt prices continued to escalate rapidly as American and European fashion houses promoted fur consumption ever more aggressively. During the 1925–1926 trapping season, the market price for individual muskrat pelts peaked at $1.30 ($17.24 in 2015 currency).

Marsh dwellers quickly adjusted their seasonal occupational calendar to accommodate this seismic shift in the marketplace, and, from late fall to early winter as many as 20,000 Louisiana *piêgeurs* spent their days inspecting muskrat traps in the coastal marshes.

Coastal trappers soon discovered, to their chagrin, that commercial trapping was far more economically volatile than market hunting had been, with peaks and valleys often occurring within the span of a single decade. The nascent Louisiana fur industry flourished until the Great Depression, when the demand for pelts plummeted. By 1933 the industry was in free fall, as the price of a muskrat pelt declined from $1.42 to $0.19 ($2.52 in 2015 currency). Even though their numbers declined in the depths of the economic crisis, Louisiana's trappers harvested an annual average of four million pelts between 1930 and 1950.

This camp on East Constance Bayou is a representative example of the numerous trappers' shanties that once were found in Louisiana's coastal marshes. Note the outhouse and two steel-barrel cisterns that constituted part of the compound. (Photo courtesy of the Louisiana Department of Wildlife and Fisheries)

An early twentieth-century Louisiana trapper processing a muskrat pelt at his camp. (Photo courtesy of the National Archives and Records Administration, Prints & Photographs Division, College Park, MD, photographer and date unknown, call number RG 22-G, Box 21, photo number 228a)

To ship nutria pelts to market, trappers put their dried skins in burlap sacks hand-stitched together. The bags were a woven, rough cloth made from jut or hemp and available in bulk from suppliers, such as the International Harvester Company New Orleans Twine Mill and Chase Bag Company. (Photo courtesy of the National Archives and Records Administration, Prints & Photographs Division, Records of the Fish and Wildlife Administration, College Park, MD, photographer and date unknown, call number RG 22-MW)

In California, New York, Florida, and Louisiana, tongs were the typical oysterman's harvesting tool of choice. Tongs consisted of twin metal heads—closely resembling bow rakes with stationary tines—each attached to long, hinged wooden handles up to twenty feet in length. The length of the handle varied according to the depth of the reef being harvested. (Most Louisiana reefs lay six to twelve feet below the surface.)

After lowering their tongs to reef beds, oystermen standing in small boats pinched the tong heads together tightly by opening the handles before lifting the captured oysters to the surface and depositing them in the vessel's hold. They tirelessly repeated this process until either their respective vessels were filled to the gunwales, darkness descended, or inclement weather forced a return to camp. After a successful outing, oystermen utilized shovels to offload oysters onto freight boats or canning company wharves.

Throughout the coastal United States, the oyster tong was the principal harvesting tool. The length of the handle is a visual indicator of a reef's depth. (Photo courtesy of the National Archives and Records Administration, Prints & Photographs Division, Department of Commerce and Labor, Children's Bureau, 1912–1913, College Park, MD, photographer Lewis W. Hine, 1909, call number LOT 7476, no. 0574 or LC-H5-574)

The sail-powered lugger was a bulwark of the early Louisiana oyster fishing fleet. The laundry drying on the rigging, whether the vessel was in Louisiana, Apalachicola, Galveston, or Biloxi, is a clear indication that the boats were the oystermen's home away from home (Photo courtesy of the National Archives and Records Administration, Prints & Photographs Division, Department of Commerce and Labor, Children's Bureau, 1912–1913, College Park, MD, photographer Lewis W. Hine, 1909, call number LOT 7476, no. 0568 or LC-USZ62-55647)

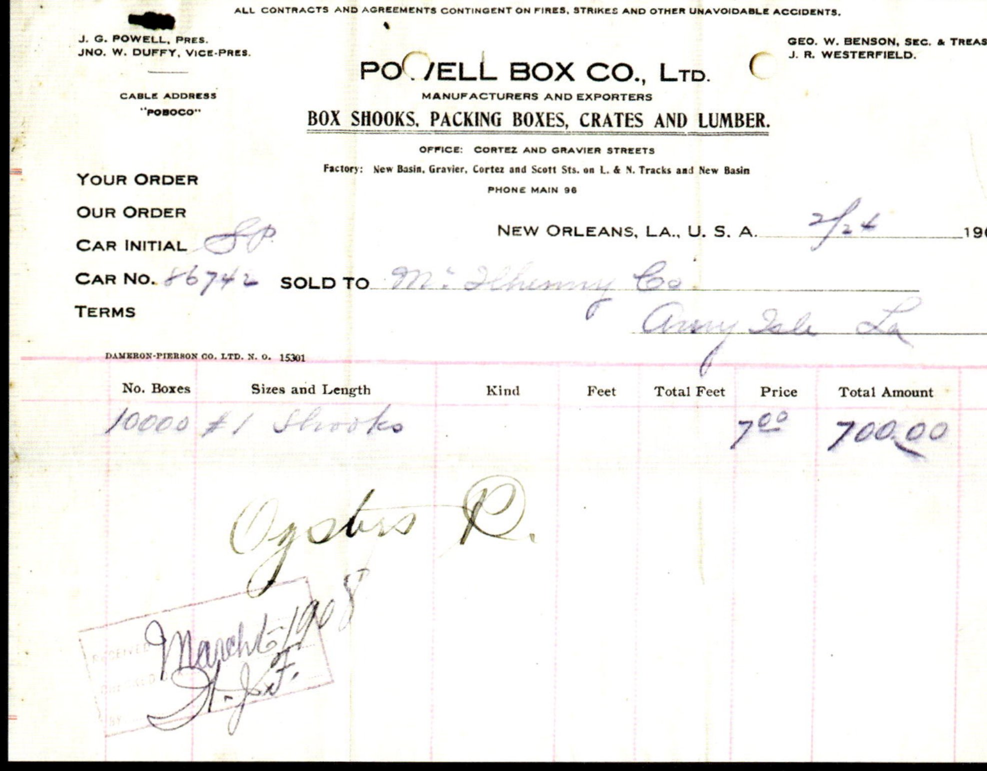

ALL CONTRACTS AND AGREEMENTS CONTINGENT ON FIRES, STRIKES AND OTHER UNAVOIDABLE ACCIDENTS.

J. G. POWELL, PRES.
JNO. W. DUFFY, VICE-PRES.

GEO. W. BENSON, SEC. & TREAS.
J. R. WESTERFIELD.

CABLE ADDRESS
"POBOCO"

POWELL BOX CO., LTD.

MANUFACTURERS AND EXPORTERS

BOX SHOOKS, PACKING BOXES, CRATES AND LUMBER.

OFFICE: CORTEZ AND GRAVIER STREETS

Factory: New Basin, Gravier, Cortez and Scott Sts. on L. & N. Tracks and New Basin

PHONE MAIN 96

YOUR ORDER
OUR ORDER
CAR INITIAL SP
CAR NO. 86742 SOLD TO Mc Ilhenny Co
TERMS Avery Isle La

NEW ORLEANS, LA., U. S. A. 2/24 190

DAMERON-PIERSON CO. LTD. N. O. 15301

No. Boxes	Sizes and Length	Kind	Feet	Total Feet	Price	Total Amount
10000	#1 Shooks				7.00	700.00

Oysters R.

March 6 1908

A 1908 invoice from the Powell Box Company to the McIlhenny Company for 10,000 shooks—the staves used in barrel construction. (Invoice courtesy of Dr. Shane K. Bernard Archivist, the McIlhenny Company, Avery Island, LA)

The fur market began to rebound by 1950, as designers such as Christian Dior, Coco Chanel, and Balenciaga introduced collections featuring casual clothing incorporating furs. In 1950 Louisiana issued 7,732 trapping licenses, reflecting the market's rebound. By the end of the 1960s, however, the sale of luxury furs again declined as popular tastes changed and faux furs emerged. The downturn in the marketplace resulted in a corresponding decline in the number of licenses issued by the Louisiana Department of Conservation. In the 1971–1972 season, the number of *piêgeurs* fell to 2,761.

Louisiana's trapping industry rebounded between 1973 and 1987 thanks to the fickle fashion industry's renewed interest in furs. By the 1990s the international fur industry began a devastating downward spiral as a result of shifting popular attitudes toward animal rights. This downturn, which was initially felt most intensely in the luxury fur market, is clearly reflected in the decline in the number of pelts taken over the last two decades. Between 1996 and 2002, Louisiana trappers annually harvested around 100,000 nutria pelts valued individually at $3.81, but by 2003 the global fur market had virtually disappeared; federal and state governments were consequently compelled to institute a bounty program to control the number of fur-bearers in the coastal plain. This program was absolutely essential because nutria, an invasive species introduced in the early twentieth century by persons with vested interests in expanding the area's fur harvest, enjoyed explosive population growth when trapping activity declined. Nutrias are voracious eaters, and if their numbers are left unchecked, they denude coastal areas of vegetation, creating, in the process, a "rotten wasteland." To help control this looming environmental Armageddon, Louisiana, acting in cooperation with its coastal restoration partners, initiated in the 2003 trapping season a $4.00 per tail "incentive payment" through the Coastwide Nutria Control Program. By 2007 the state had paid more than $6 million in bounties.

In recent years the moribund Louisiana fur industry's fortunes have brightened somewhat as a result of shifts in the global marketplace. In North America and Europe, where eco-conscious millennials have begun to value recycled vintage furs, demand for new furs remains essentially nonexistent. Instead, the industry's hope in the twenty-first century lies in the Pacific Rim, where the demand for North American pelts is expected to intensify. However, in the coastal plain, the fur industry's future is highly problematic. As a result of the disruptions caused by the effective antifur campaign of the recent past, the critical transmission of interest and practical skills from one generation to another has been broken, and, in the interim, the youth in many coastal communities have learned to supplement their income by other means.

We Own and Control Our Trapping Lands

Long Distance Phone MAin 6023

Martinez-Mahler Company

RAW FURS

221 South Peters Street

NEW ORLEANS, LA.

As the demand for muskrat pelts increased in the early twentieth century, New Orleans buyers forged links between Louisiana trappers and New York couturiers. (From the authors' collections)

Labor unrest in the trapping industry. The Southern Fur Trappers Union organized these picketers against the Steinberg Company. (Photo courtesy of the Louisiana Department of Wildlife and Fisheries)

The Dunbar Molasses and Syrup Company factory at 333 Chartres Street, New Orleans, has been repurposed as the headquarters for Turn Administrations. (Photo by the authors, 2014)

The Dunbar Company and its rivals sold their products under a variety of trade names and logos in never-ending struggles to dominate national and regional markets. (From the authors' collections)

Shrimp have been described as the most popular shellfish in the United States. This was not always the case, because the crustacean's short shelf life limited its consumption to coastal markets. Canning changed this paradigm, and canners quickly learned that publication and distribution of free cookbooks were essential to enticing consumers in previously untapped markets to sample their wares. (From the authors' collections)

..1865..

ORIGINAL DUNBAR SHRIMP

Are sufficiently cooked to be eaten cold or prepared in countless ways.

Ask your grocer. He has known this package for more than fifty years, and will tell you that ORIGINAL DUNBAR SHRIMP is always reliable, uniform and dependable, and is backed by a reputation of more than fifty years.

Try a can to-day.

PACKED BY
DUNBAR-DUKATE CO.
NEW ORLEANS, LA.
MORE THAN FIFTY YEARS.

Brochures like the one pictured here helped convince wary US consumers that canned shrimp were "sufficiently cooked to be eaten cold or prepared in countless way." (From the authors' collections)

After the decline of sail power and the introduction of maritime gas engines, the canning industry established, coordinated, and sustained fleets of shrimp and oyster boats. (Photo courtesy of the National Archives and Records Administration, Prints & Photographs Division, College Park, MD, photographer Ben Shahn, 1935, call number LC-USF33-002056-M5, 8a07130u)

During the early post–Civil War era, entrepreneurs introduced into Louisiana technological innovations that helped make the state a national leader in seafood production, processing, and distribution. In 1865 George H. Dunbar's company began canning a multitude of semitropical products. (This operation predated by four decades the legendary "Cannery Row" in Monterey, California.) Dunbar's product lines enjoyed consistently strong regional sales, while its "Barataria" shrimp brand was a national brand leader thanks to its enduring cachet among mainstream seafood consumers. By 1885 the Louisiana cannery had established a thriving export business, with distribution networks in England, France, Germany, Spain, and Mexico.

In the 1930s Morgan City made national headlines by promoting itself as the "jumbo" shrimp capital of the world, and the oxymoron "jumbo shrimp" entered the national lexicon. (Photo courtesy of the Morgan City Archives, Morgan City, LA, photographer and date unknown)

As canning machinery became more readily available, packers throughout Louisiana's wetland communities began to get involved in the industry. This label from the Lafourche Packing Company in Golden Meadow identifies one participant. (From the authors' collections)

The first day of Louisiana's September alligator season can be highly profitable for trappers. (Photo by the authors, 2013)

Not all trapping activity appears to be doomed to oblivion. To the contrary, the regional alligator industry remains robust and profitable. Louisianians have hunted alligators for their meat and hides since colonial times. Extant documents, however, strongly suggest the industrialization of this pursuit did not begin in earnest until the mid-to-late nineteenth century, particularly after 1855, when Parisian fashion designs made alligator leather chic.

The commercial alligator hunting industry was thus only in its infancy when the Civil War intervened. A Union blockade at the beginning of the war disrupted the export of hides to external markets. As a result, the alligator trade appears to have been limited to military and domestic consumption, particularly during the very early stages of the conflict. Nevertheless, as the geographer Gay Gomez correctly observes, "by 1870, alligator leather was once again the height of fashion, and increased prices sent hunters into the marshes in search of 'alligator gold.'"

The postbellum commercial harvesting of alligators began and continued for some time at a rather modest scale. In 1889, for example, official federal documentation indicates that 169 Louisiana hunters sent to market 40,608 hides. (To put this figure into proper perspective, Florida's 1,207 commercial alligator hunters produced 84,938 skins that year.) Federal officials expressed grave concern that, as with the plumage trade, the existing scale of alligator carnage was not sustainable. Nevertheless, during the Great Depression, the Louisiana conservationist E. A. McIlhenny estimated that "3 million alligator[s] were killed in the state between 1880 and 1933"—an average of approximately 56,600 per year. The ongoing slaughter and habitat loss resulting from sustained wetland modifications in the late nineteenth and early twentieth centuries resulted in a steadily declining alligator population. Between 1939 and 1960, official state reports indicate that harvested skins plummeted to approximately 18,005 per year, a level not seen since the early post–Civil War era.

Decimation of the alligator population compelled Louisiana to adopt stringent conservation measures. In 1960 the state established a brief hunting season and imposed size limits, and two years later a hunting moratorium was introduced. Five years later *Alligator mississippiensis* was listed as an endangered species under conservation legislation predating the Endangered Species Act of 1973. Thanks to the proscription, Louisiana's alligator population rebounded rapidly. On September 5, 1972, Louisiana officials reopened the hunting season in Cameron Parish, and fifty-nine hunters harvested 1,350 alligators. Other parishes were systematically added to the September controlled hunt. By 1981 tightly controlled alligator hunting was legal statewide.

The ready availability of cotton twine and local expertise made net-making an important industry in the New Orleans area. (Lithograph from *Harper's Supplement*, September 1, 1888, 78)

Shrimping was initially a cottage industry in Louisiana. Local residents caught freshwater prawns in the Mississippi River and transported them home with homemade baskets for domestic consumption. (Lithograph from *Hearth and Home Magazine,* April 6, 1872)

Tiny freshwater prawns—known locally as *crevettes de fleuve*—were common in the Mississippi River until around 1960. This prawn variety (*Macrobrachium ohione*) is also called "Caridean" shrimp. The historic record notes river shrimp were present in the days of the early colonization of the Mississippi floodplain, and they were an important food source for poor New Orleanians. Often a cook would lower a barrel, baited with table scraps, to just below the water's surface and after two or three days raise it to harvest a generous serving of "shrimp." (Lithograph from "Southern Scenes, Fishing on the Mississippi River," *Frank Leslie's Popular Monthly* [1878], nonpaginated)

The world-famous Tabasco brand pepper sauce, manufactured by the McIlhenny Company on Avery Island (previously known as Petite Anse Island) in Iberia Parish, is perhaps Louisiana's signature hot sauce. The pepper *Capsicum frutescens* var. *tabasco* first appeared on the island in 1849 or 1850, when Edmund McIlhenny planted seeds provided by a friend, Colonel Maunsell White, who had recently returned from Mexico. McIlhenny experimented with the seeds and plants until the outbreak of the Civil War. During the conflict, the McIlhenny family fled the island to escape an imminent Union invasion and took refuge in San Antonio, Texas. Upon returning to their former plantation home, Edmund found his plantation devastated and the crops destroyed. However, his cherished pepper garden survived, providing the seeds for McIlhenny to continue experimenting with his Tabasco peppers, largely in an effort to restore the family's flagging economic fortunes. By 1868 he had devised a satisfactory hot sauce formula, and he sold 350 bottles of the condiment—bearing the Petite Anse Island Sauce label—to a wholesale grocery house. This sale launched one of Louisiana's most famous family-owned industrial enterprises. In 1870 McIlhenny obtained a patent for his Tabasco sauce, and in 1906 his descendants trademarked the name Tabasco to distinguish it from emerging imitators seeking to trade on the established product's name recognition and profitability.

Like all of Louisiana's major canning firms, the McIlhenny Company issued products under a surprising assortment of labels. To take advantage of the Chesapeake Bay area's marketing cachet, this label indicates that it was a product of the "Baltimore Packing Company," but it concedes, in fine print, that the cans were "Packed at Avery Island, Louisiana." (Label courtesy of Dr. Shane K. Bernard Archivist, the McIlhenny Company, Avery Island, LA, date unknown)

Tabasco's success stemmed not only from the product's inherent quality, but also from the company's marketing acumen. The condiment quickly permeated the national markets, and in 1872 McIlhenny opened an office in London to oversee the condiment's European distribution. The production and processing of the product, however, remained at Avery Island, where, between 1869 and 1909, the hot sauce was bottled and packaged in a building known simply as the "laboratory."

The industrial production and distribution of Tabasco sauce has remained at Avery Island to the present, but the company's agricultural arm has, of necessity, migrated in large part to Central and South American farms, where manual labor is more easily procured. Barrels of pepper mash from these farms are transported to Avery Island for aging and bottling.

CHAS. SCHULER, Commissioner.

Louisiana State Board of Agriculture and Immigration.

Immigration Division,
730 Carondelet Street.
J. L. KNOEPFLER, Secretary.

NEW ORLEANS, LA., May 13th, 1907.

Mr. E. Mc.Ilhenny's Son,

New Iberia, La.

Dear Sir:-

Your letter of the 8th inst has been received, and in reply to same would say that during our trip in Europe we will try to induce some families to come and settle in your country. The conditions you express are very good, and we expect some of the families will take advantage of your offer.

Respectfully,

Chas. Schuler, Commissioner,
Per. ______
Secretary,

Labor issues after the Civil War led the Louisiana State Board of Agriculture and Immigration to mount a recruitment campaign for replacement workers. Although the sugarcane industry was the primary beneficiary of the new recruits, other industries were also interested in hiring new immigrants. (Letter courtesy of Dr. Shane K. Bernard Archivist, the McIlhenny Company, Avery Island, LA, 1907)

LABOR CONTRACT.

THE McILHENNY CANNING & MANUFACTURING CO., of New Orleans, La., herein represented by E. A. McIlhenny hereunto duly authorized, hereinafter known as PARTY OF THE FIRST PART, and, Louis Williams & Bessy Williams his wife & John Williams his brother.

herein known as PART 3 OF THE SECOND PART, enter into the following LABOR CONTRACT, in accordance with the provisions of act No. 50 of the Legislature of the State of Louisiana for the year 1894, upon the following condition to-wit:—

THE PART 3 OF THE SECOND PART agree to enter the employ of and engage their service to the PARTY OF THE FIRST PART on the following conditions to-wit:—

The time of employment shall be for the period of the oyster season of the PARTY OF THE FIRST PART.

The rate of compensation on salary shall be ______

And the time of payment shall be ______

The character of service rendered shall be the men as boatmen or tongmen. The women as shuckers

The services shall be rendered at, in, or for, the Canning Factory of the PARTY OF THE FIRST PART located at AVERY'S ISLAND, PARISH OF IBERIA, STATE OF LOUISIANA, where PARTY OF THE FIRST PART in engaged in the business of Canning and Packing OYSTERS, SHRIMPS, VEGITABLES and FRUITS.

This LABOR CONTRACT is to be executed, and the services to be performed in the State of Louisiana and is to be construed according to the Laws of that State.

For and in consideration of the foregoing premises and upon the faith thereof and of this Contract, the said PARTY OF THE FIRST PART accepts the conditions of this LABOR CONTRACT agrees to furnish the employment and pay the wages aforesaid and agrees to advance to the said PART ______ OF THE SECOND PART transportation from ex Baltimore to AVERY'S ISLAND, LA., and upon the compliance by PART ______ OF THE SECOND PART with the terms of this contract, to furnish free return transportation, at the end of this Contract, from AVERY'S ISLAND, LA., to Baltimore

Additional stipulations:

THUS DONE AND SIGNED in duplicate this Oct day of 15 1906.

McILHENNY CANNING & MFG., CO.,
By E. A. McIlhenny
Louis Williams
Mrs Bessie Williams
John Williams

A 1906 McIlhenny Canning and Manufacturing Company labor contract for "Bohemians" employed in its oyster facility. (Agreement courtesy of Dr. Shane K. Bernard Archivist, the McIlhenny Company, Avery Island, LA, 1906)

Since colonial times, oysters from Louisiana's wetlands have sustained the regional population. (Photo courtesy of the National Archives and Records Administration, Prints & Photographs Division, College Park, MD, photographer Russell Lee, 1938, call number LC-USF33-01-763-M2, 8a23976a)

A wooden oyster lugger, outfitted with an auxiliary sail, moving a fisherman's harvest to market. (Photo courtesy of the Louisiana Department of Wildlife and Fisheries)

Thirty years after the first monthlong harvest, 2,595 commercial hunters tagged 35,760 alligators. Louisiana's Department of Wildlife and Fisheries alligator management program has successfully resurrected the American alligator as a commercial, renewable natural resource. Between 1972 and 2014, more than 870,000 wild alligators have been tagged, skinned, and processed for their meat and hides.

Strict resource management was also necessary to achieve sustainability in the regional seafood industry. In the industry's infancy, governmental regulation was unnecessary because it serviced small, frequently economically depressed local markets. In the late eighteenth century, New Orleans's open-stall public markets constituted Louisiana's principal commercial outlets for local fin- and shellfish. The scanty relevant contemporary documentation suggests that oysters dominated the regional shellfish trade. The bivalves evidently found a ready market in the Crescent City, particularly after the Louisiana Purchase (1803), when newspaper advertisements document the emergence and proliferation of numerous oyster saloons—all within a short distance of the Carondelet Canal (also known as the Old Basin Canal) connecting New Orleans's commercial district with nearby coastal reefs between 1794 and the late 1920s. Over time, oyster boats emanating from reefs west of the Mississippi came to utilize the wharves—including the Governor Nicholls Street Wharf—along the Crescent City's riverfront.

The geographical reach of Louisiana's early commercial fishing industry would remain fixed until technological innovations permitted more expeditious transportation of highly perishable seafood. By 1850 the Mexican Gulf Railroad extended from downtown New Orleans to Shell Beach in St. Bernard Parish, permitting the transportation of oysters to urban markets within hours. At the Governor Nicholls dock, laborers packed fresh oysters into barrels and rolled them directly onto boxcars.

A second railway, the New Orleans, Opelousas, and Great Western Railroad, which extended from the Crescent City to Brashear (now Morgan City) in 1857, theoretically tapped new sources of supply in the southeastern and south-central portions of the coastal plain. However, the travel times incurred by oystermen in moving product from coastal reefs to inland depots made this trip impracticable.

Oyster fishermen living beyond the effective reach of the railroads had to transport their catch to New Orleans by sailboats. Whenever their luggers were becalmed, oystermen were compelled to return their catch to water and wait for at least two tides before retonging them and continuing the arduous journey. The transportation of oysters to market from outlying regions was not only grueling but also extremely costly to fishermen, who received no income while the bivalves were in transit.

A revolution in food preservation technology in the late nineteenth century afforded oystermen not only a more efficient procedure for moving their "harvest" to market but also a method of reaching an exponentially larger market for their bivalves. Gulf Coast residents had traditionally relied upon salting, smoking, drying, and hermetic sealing (that is, crock pots with lard seals) to preserve perishables. However, none of these techniques was suitable for oysters prized for their freshness and unadulterated salty taste. Canning afforded oystermen the first viable technology for preserving the oyster's raw flavor while drastically increasing its shelf life.

Canning technology was a direct outgrowth of the French military's effort in the mid-1790s to secure a new food preservation method to support its armies in the field. The first American canning facility was built in New York in 1812. Utilizing tin-plated cylindrical iron containers (quickly identified in the vernacular as "tin cans"), American canning factories came into their own during the Civil War. In the wake of the conflict, commercial canners diverted their expanded wartime production capacity to the civilian market, and by 1880 canned products became widely available commercially, at least to the American middle class.

This shift coincided with the metamorphosis of the national oyster industry's infrastructure. New York, the epicenter of the nation's oyster trade in the early nineteenth century, declined in importance over the antebellum era as a result of overharvesting. By the post–Civil War era, Chesapeake Bay emerged as the national leader in oyster production, but, like its northern neighbor, it, too, succumbed to overfishing over the course of a half-century. As the Chesapeake's output declined, the industry's focus shifted to the Mississippi delta, where highly productive natural reefs, supplemented by artificial beds seeded by recent Croatian immigrants, quickly outstripped their East Coast counterparts.

As its focus shifted to the upper Gulf Coast, the industry became largely dependent upon the success of ancillary canning operations. During its heyday, the Chesapeake oyster canning industry was based in Baltimore, whose famous "cove" oysters (named after the lane where several shucking houses were located) constituted the national benchmark for excellence. The Baltimore canneries, which boasted national distribution networks by century's end, were staffed by a seasonal workforce, consisting primarily of eastern European immigrants popularly known as "Bohemians."

As oyster production shifted to the Gulf Coast, so did the industry's attendant canning industry. By 1920 eleven of the nation's sixty-five oyster canneries were located in New Orleans, while some of the world's largest seafood canneries were located in Houma and along the Mississippi Gulf Coast. Like all successful industrial operations,

In 1859 Frenchman Ferdinand Carré invented the first practical modern refrigeration system. The device was patented the following year. Carré's invention proved a godsend for New Orleans, which had been dependent upon ice shipments from northern states—particularly the Northeast—throughout the antebellum era for its refrigeration needs. Trade disruptions during the Civil War and lingering regional animosities following the conflict dictated that the Crescent City seek another source of supply. By the end of 1865, New Orleans boasted three of Ferdinand Carré's ice machines.

Before the Civil War, ice was a luxury in Louisiana's coastal plain. The introduction of ice factories in the postbellum era and the proliferation of the industry in the late nineteenth and early twentieth centuries made ice an essential commodity in urban and small-town Louisiana. (Photo courtesy of the National Archives and Records Administration, Prints & Photographs Division, College Park, MD, photographer Russell Lee, 1938, call number LC-USF33-011712-M3, 8a23976a)

Living in isolated villages built on stilts around large drying platforms (some as large as three acres), significant numbers of undocumented Chinese laborers—many of whom had been shanghaied and forced into a state of involuntary servitude, maintained by the constant threat of physical violence—dried fresh shrimp on gently undulating decks and then packed the desiccated crustaceans in barrels for export. This largely forgotten chapter in coastal Louisiana's economic development is of vital historical importance to the state's seafood industry, because it opened the door to international exports. Dried shrimp were first exported to China, but by the turn of the twentieth century the industry's export market had expanded to include the New Orleans–based United Fruit Company's immense Central American banana plantations.

The Chinese shrimp drying platforms proved unsustainable, not because of the harvesting and preservation methods used by their residents but because of the sites' vulnerability to hurricane tidal surges. Anecdotal information gleaned from oral histories indicates that entire communities drowned during the hurricane of 1915. A 1917 report to the US Congress by Frederic V. Abbot suggests that the shrimp-drying industry recovered to some extent from this natural disaster, but, when the most exposed platforms were destroyed—again with heavy loss of life—during the 1926 storm, many owners chose to cease operations.

In 1882 the Chinese Exclusion Act (from 1882 to 1943—sixty-one years) prohibited both Chinese immigration as well as Chinese immigrant naturalization. New Orleans nevertheless had a small, but thriving, Chinese community. (Photo courtesy of the National Archives at San Francisco, Immigration Files, San Bruno, CA, date unknown)

Asian laborers operating shrimp-drying platforms grew their own vegetables on-site. They used "night soil" packed in wooden boxes to aid in producing the vegetables they needed for their wok-based cooking. (Photo courtesy of the Louisiana Department of Wildlife and Fisheries)

The Blum and Bergeron Company of Houma once enjoyed a monopoly on dried shrimp exports to C (left). Further, in 2015 the Obama administration m to normalize relations with Cuba. This move was seen as a boon for the Louisiana rice industry, whi dominated the Cuban rice market before 1959 and a major destination for Louisiana dried shrimp (ab

This photo, probably one of the earliest industry images on record, clearly indicates that the art of preparing fresh shrimp for drying was a time-consuming, labor-intensive undertaking. (Photo from Tate Album, Mss. 4963, LLMVC, photo number 4963064)

Small seine crews harvested estuarine shrimp at remote small sites, where they deployed bags on the marsh surface to sort and dry the crustaceans. (Photo from Tate Album, Mss. 4963, LLMVC, photo number 4963063)

The structure pictured here represents an intermediate stage of development in the evolution of dried shrimp platforms. Workers used wooden barrels to package the dried shrimp. The device in the center of the image was apparently a homemade tool that allowed the fishermen to sift the shrimp and remove the outer hull. Readers should keep in mind that everything in this image had to be delivered to this site by boat. (Photo from Tate Album, Mss. 4963, LLMVC, photo number 4963065)

Folk boats were the original workhorses of Louisiana's seafood industry. Marsh dwellers used *esquifs* and red-sail luggers, outfitted with haul seines and a crew of up to twenty to manage fishing nets. In the first quarter of the twentieth century, these sail-powered craft were gradually replaced by vessels with gasoline-powered motors and propellers. The new boats boasted better handling and maneuverability than their predecessors. During the fishing fleet's conversion to gas power, a process nearly complete by the 1930s, two new boat types predominated. The first, known locally as the "*Floridiane*," featured a forward wheelhouse, while the second, known colloquially as the "Biloxi," boasted a prominent wheelhouse in the stern. Although these innovative boats had sufficient power to pull nets through the inland waters, they lacked both the ability to produce ice and the speed necessary to move their catch quickly to the dock. The system's inherent limitations were compounded by the fact that state-sanctioned seasons were short (May–June and August–November) and that a time-consuming return to port severely hampered a shrimper's ability to generate a profit. In addition, the absence of onboard ice machines significantly heightened the possibility that a fisherman's highly perishable catch would be rancid upon arrival at the docks. These realities dictated the creation of a new logistical support system for the state's shrimp fleet.

The roots of this new system can be traced to efforts by local fishermen to organize into "companies" or syndicates. Each syndicate designed a flag identifying participant boats. In the Bayou Lafourche area, there were at least fifteen such companies. Each of these enterprises operated three relatively large watercraft, dubbed "mother ships" by some locals, "freight boats" by others. These vessels functioned as floating refrigeration, logistical support, and resupply units, into which shrimpers funneled their catch.

The three mother ships typically sailed together to a predetermined rendezvous site in an estuary serviced by the syndicate's shrimp boats. One of these mother vessels was empty, while the others were loaded to the gunwales with ice and fuel. At the rendezvous site, shrimpers' relatively small trawlers would successively offload their catch into the initially empty transfer vessel. A second mother ship then provided ice to the boat receiving the transferred shrimp. When fully laden, the vessel transporting the shrimp made the five-to-ten-hour trip to a shrimp shed or cannery, where the offloading process took about twenty-four hours. Once the offloaded shrimp were sold and the transfer boat's hold was loaded with a fresh cargo of ice and fuel, the vessel returned to the harvest site, where it awaited its turn to take on a fresh cargo of shrimp. Meanwhile, following the first vessel's departure from the estuary, the second, now empty mother boat assumed the role of shrimp receptacle, guaranteeing a seamless harvesting and processing operation.

The mother ships also provided logistical support to syndicate boats in the estuaries. In order to operate in coastal waters as long as possible, trawlers were periodically refueled by the mother boats. This logistical support allowed shrimpers to maximize profits by trawling for longer periods. In fact, it was not uncommon for trawlers to operate continuously for four to six weeks at a time.

A flotilla of oyster luggers tied up at a processing plant. (Photo courtesy of the National Archives and Records Administration, Prints & Photographs Division, College Park, MD, photographer and date unknown, call number unknown)

Fishermen are seen here "seeding" oyster beds. (Photo courtesy of the National Archives and Records Administration, Prints & Photographs Division, College Park, MD, photographer and date unknown, call number unknown)

A day laborer shoveling oyster shells at a Berwick processing plant. (Photo courtesy of the National Archives and Records Administration, Prints & Photographs Division, College Park, MD, photographer and date unknown, call number unknown)

MANUEL COGUENHEM, - President.
LEON BLUM, - - Vice-President.
MAURICE BLUM, Secretary-Treasurer.

BERWICK BAY
FISH & OYSTER CO.

LIMITED

...WHOLESALE AND RETAIL SHIPPERS OF...
The Celebrated Berwick Bay
Oysters and Cat Fish

And Dealers in General Merchandise

Morgan City, - - Louisiana.

The Morgan City area constituted an early seafood industry nexus outside New Orleans. The urban communities in this area fortuitously sat at the intersection of the Southern Pacific Railroad, which provided quick and reliable access to the New Orleans and Houston markets, and the Atchafalaya River, which afforded fishermen direct access to local sources of freshwater and saltwater fin- and shellfish in the Atchafalaya Basin and Atchafalaya Bay. (Advertisement courtesy of the Morgan City Archives, Morgan City, LA, date unknown)

Daspit, a now largely forgotten coastal community, served as one of Terrebonne Parish's principal oyster lugger bases. (Map from Rand McNally's *Indexed Atlas of the World, Louisiana* [Chicago: Rand, McNally, 1892])

The oyster lugger *Detroit* clearing the locks at Empire. (Photo courtesy of the Louisiana Department of Wildlife and Fisheries)

In 1911 the pioneer American sociologist Lewis W. Hine reported that "a line of canning factories stretches along the Gulf Coast from Florida to Louisiana. After many weeks spent in those canneries, I am convinced that, individually and as a 'great and glorious people,' we ought to be ashamed of ourselves. I have witnessed many varieties of child labor horrors from Maine to Texas, but the climax, the logical conclusion of the 'laissez faire' policy regarding the exploitation of children is to be seen among the oyster-shucks and shrimp-pickers of that locality."

Source: L. W. Hine, "Child Labor in Gulf Coast Canneries: Photographic Investigation Made February, 1911," *Annals of the American Academy of Political and Social Science* 38 (1 suppl.) (1911): 118–122.

Fisher Shrimp Co., Inc.

Plants
CABINASH, LA.
GRAND ISLE, LA.

PACKERS OF SUN-DRIED SHRIMP

Postoffice: Cabinash, La.
New Orleans Office: 822 Perdido St.

Dried shrimp dealers were some of southeastern Louisiana wetland's first successful entrepreneurs. (From the authors' collections)

these seafood factories required an uninterrupted source of supply, which meant that canneries and oystermen required a far more efficient means of transporting freshly tonged bivalves from remote oyster leases to processors. The introduction of gasoline-powered internal combustion engines in the early twentieth century solved this once seemingly intractable problem. This "new" technology made it possible for major canneries to secure fresh oysters directly from distant coastal reefs. Gulf Coast oystermen typically resided seasonally at these isolated reefs to protect their beds from predation by competing humans and fauna (for example, drum and oyster drills).

Such on-site security resulted in impressive dividends. By 1912 Louisiana produced 2.3 million bushels of oysters worth approximately $1 million ($24,030,303 in 2014 currency). Because oystermen were tied to their leases, the most enterprising and visionary cannery entrepreneurs moved their facilities as close as possible to the source of supply to reduce the expense and time lost in transporting bivalves. Entire canneries and their ancillary buildings—usually a company store, workers' dormitories, and privies—eventually were mounted on barges and floated to remote locations.

The Dunbar family of New Orleans pioneered the use of these portable factories. Shortly after the outbreak of the Civil War, businessman George W. Dunbar sent his sons Frank B. and George H. to France, where they studied fruit syrup manufacturing and food preservation techniques. Upon returning to New Orleans in 1867, the brothers established George H. Dunbar & Sons Cannery to process such semitropical local products as shrimp, green turtles, figs, oranges, okra, and myriad other fruits and vegetables. Located on Chartres Street, between Piety and Desire Streets, the Dunbar operation, which shipped large quantities of canned foods to northern states as well as to Europe and Mexico, established New Orleans as the epicenter of the South's food preservation industry.

Over the succeeding decades, the proliferation of canneries in seafood-rich southeastern Louisiana forced Dunbar & Sons to alter its business model to maintain market share. The company first established a foothold in Biloxi, Mississippi, which was rapidly emerging as a major American seafood processing hub—to which much Louisiana seafood was being diverted. In Biloxi, the Dunbar Corporation merged with a similar enterprise owned by Lazaro Lopez to create what developed into the nation's second-largest oyster cannery by 1895. (The factory also canned shrimp.) The facility was supplied by forty-five boats, serviced by more than 100 shuckers.

In September 1908 Lopez, Dunbar and Sons merged with the DuKate Company, founded by W. K. M. DuKate, who learned the canning business in Baltimore. By 1908 the consolidated operation was America's largest canning company, with more than 200 vessels operating along the Louisiana and Mississippi coasts. Most of the "catch" aboard these boats, however, emanated from Louisiana waters.

Louisiana could not afford to ignore the migration of numerous jobs and considerable tax income to its eastern neighbor, and in 1910 the state's general assembly adopted the Oyster Law of the State of Louisiana, mandating that "no person, firm or corporation shall ship oysters out of this state for canning or packing out of this state." Dunbar, Lopez, and DuKate responded by establishing factories at Dunbar, sometimes called Lookout; just west of the Louisiana-Mississippi line; at Neptune, on the east bank of the Mississippi in Plaquemines Parish, near the Ostrica oyster community; and at the Rigolets. The firm later established canning operations at Morgan City, Empire, and Myrtle Grove, while also constructing a major west-bank seafood wharf serviced by all of the region's major railroads.

The company's most notable achievement, however, was the introduction of floating factories designed to locate canneries as close as possible to often remote sources of supply. The floating factories' workforce, consisting of up to 300 laborers, primarily Bohemians who generally did not speak English, endured harsh working conditions. Fourteen-hour workdays were the rule. At land-based sites the transient laborers—adults *and* children—occupied cramped dormitory rooms, measuring only ten by five-and-a-half feet—an area slightly smaller than a modern metal shipping container. These spaces were crammed with bunks, a wood-burning stove, and a primitive table and bench, and the walls were usually so porous that workers had to plaster their walls with foreign-language newspapers to keep the winter winds at bay.

As processing plants and canneries sprang into existence, Gulf Coast entrepreneurs not only modeled their industrial infrastructure on that of Maryland, but they also utilized Baltimore labor brokers to provide capable, seasonal workers. Every year trainloads of experienced "Bohemians" debarked at canneries from Avery Island, Louisiana, to Biloxi, Mississippi. In these factories, imported workers—including a large number of women and children—endured working and living conditions at least as harsh as those they experienced in Baltimore. Indeed, these deplorable conditions prompted New Orleans socialites Jean and Kate Gordon to initiate one of the nation's first—and arguably most successful—child labor reform movements.

Other Maryland influences were more ephemeral. Louisiana factories initially sought to capitalize upon cachet surrounding the trade term Baltimore "cove," a term American consumers then associated with quality. At the turn of the twentieth century, South Louisiana and southern Mississippi cannery labels—all printed in New Orleans with brightly colored, eye-catching graphics—consistently featured the term

The American Can Company's location in New Orleans stands as an enduring reminder of the importance of the canning business to Louisiana's early seafood and produce providers. (Photo by the authors, 2011)

G. W. DUNBAR'S SONS,
NEW ORLEANS, La., U. S. A.,
The Original Packers of Fresh
BARATARIA SHRIMP,
IN LINED CANS,
Have Obtained a
SILVER MEDAL,
The only Award for Shrimp granted by the
INTERNATIONAL FISHERIES EXHIBITION,

As the canning industry matured, canners began to utilize the term "Barataria Shrimp" as a hallmark for quality. (From the authors' collections)

"cove" prominently in the brand name or generic product description as a means of duping unwary consumers into buying an ostensibly familiar, albeit untested, off-brand commodity.

Finally, upper Gulf Coast canners modeled their respective distribution networks upon those of their East Coast competitors. Growing Gulf Coast brand recognition and rapid market acceptance fueled a spiraling demand for Louisiana canned oysters, which, thanks to America's maturing rail network, were being sold nationwide. By 1913 Louisiana canned oysters were featured prominently in advertisements in the *El Paso (Texas) Herald*, while five years later, "sanitary sealed" Louisiana canned oysters were showcased in a Bisbee, Arizona, advertisement.

National distribution presented two major early challenges to the state's canning industry. First, Louisiana manufacturers had to reassure consumers that their cans were free of lead solder. This toxic substance—once widely used as a tin-can sealant—posed a significant health risk, as consumers stricken with lead poisoning could readily attest. Second, canned oysters lacked the tang of their fresh counterparts, and prolonged shelf life only accentuated the bivalves' blandness. Similarly, fresh oysters marketed by rail tended to become rancid before reaching destinations in the Great Plains and on the West Coast. Both circumstances catapulted a new Louisiana condiment—Tabasco—into national prominence. Bottled at Avery Island in Iberia Parish, Tabasco provided backcountry oyster aficionados a means of masking or counteracting the flavor of oysters of dubious palatability.

The ability to make oysters palatable to consumers in distant inland markets helped sustain the Louisiana oyster industry despite the volatility of the national economy—which experienced approximately ten major downturns between 1890 and 1920, when the state's oyster output was valued at $1.5 million. The American Can Company, a member of the nation's monopolistic Tin Can Trust, established a factory along Bayou St. John in 1908. Nine years later, the Continental Can Company opened New Orleans's second major canning facility. These facilities functioned not only as full-service canneries but also as distribution centers for licensed canning equipment, pressure cookers, and cans utilized by small canneries in predominantly rural areas.

This canning infrastructure and its attendant distribution network provided the impetus for the rapid development of the region's other notable shellfish industries—shrimp and crab processing and distribution. Development of the commercial shrimp industry was impeded in lower Louisiana by the ready availability of freshwater prawns, a staple in the diet of poor river-parish residents. Prawns were also particularly important to the diet of the Crescent City's indigent residents, who caught large quantities of the shellfish with crude nets along the riverbanks. Gulf shrimp, on the other hand, were a highly problematic commodity, because, in nineteenth-century New Orleans, they could not compete with their freshwater cousins on the basis of price or availability. Shrimp were highly perishable, and, unlike oysters, fishermen could not resuscitate their catch by returning them to water. Following the Civil War, Chinese immigrants introduced a centuries-old drying process that preserved shrimp for up to three years. The vast majority of this product, however, was transported by rail to North America's Chinatowns and to California for transshipment to China.

Like oysters, shrimp were catapulted into American economic prominence by mid-nineteenth-century technological innovations. Railroads, of course, provided direct access to distant markets, and in 1865 the establishment of the nation's first commercial ice factory made it possible to move the product from the coastal waters to South Louisiana's urban markets, where they became increasingly desirable as freshwater prawn stocks declined due to river pollution. Much more important, ice permitted fishermen to preserve their catch for delivery to canneries, where they could be processed for shipment to distant external markets.

Industrial canning operations were contingent upon an uninterrupted supply of product. In the upper Gulf Coast's seafood industry, this required not only special field operations but also specialized machinery aboard supply/product-transfer boats, colloquially known as "mother boats." When mother boats returned to the canneries, workers lowered steel baskets into their holds, where sailors filled them with up to 100 pounds of shrimp. Filled baskets were then lifted to wharves, where longshoremen transferred the contents to small flat cars with half-ton capacities. Once a car was full, workers pushed it along a narrow-gauge railway to a temporary storage facility, where the crustaceans were covered with crushed ice until they could be processed.

From the storage facility, shrimp were transferred to partitioned "picking" tables, where women and children peeled them since there were no mechanical shrimp peelers. Most workers received five cents per six pounds of peeled shrimp, and the typical "picker" earned $1.50 a day. (Highly skilled workers earned as much as $2.50 a day.) At most processing plants, pickers received payment in the form of tokens redeemable only at the factory store.

Upon completion of the "picking" process, workers weighed the shrimp and washed them in a double-walled tank. The tank's internal basket—made of galvanized

The United States' first recorded commercial crawfish harvest took place in Louisiana's Atchafalaya Basin during the 1880s. Extant documentation indicates that basin fishermen used baited lines to gather 23,400 pounds with a market value of $2,140—$0.11 a pound. In 1908 the US Census Bureau reported that Louisiana's crawfish production had increased to 88,000 pounds, valued at $3,600—$0.24 a pound. By the 1940s the catch was estimated to be in excess of two million pounds with a value of $175,000.

Crawfish boats, powered by air-cooled engines and equipped with a hydraulically driven front wheel, are extensively used by fishermen to harvest orderly rows of crawfish traps in flooded rice fields. (Photo by the authors, 2015)

metal and filled with shrimp—fit snugly inside an outer wall made of cypress. Once the meat was dumped into the washing tank, water was piped into the container, and the shrimp were subjected to two baths. The meat was then transferred to the cypress container, and the process was repeated.

Once the cleaning phase was complete, cannery laborers moved the peeled shrimp to baskets for cooking by boiling in large vats. Workers subsequently transferred the cooked meat to cans. Filled cans migrated via conveyor belts to weighing stations, then into brining machines where the cans were filled with saltwater, and, finally, sanitary sealing devices positioned solder-free caps on the containers. After the capping process, employees moved completed cans to storage areas, where, after a week, they were inspected for leaks, labeled, and boxed in wooden crates. These assembly lines were typically capable of producing up to 20,000 cans per day.

Boxed cans were shipped by rail throughout the country. Utilization of rail transportation worked to the advantage of the canneries because of the railroad industry's marked reluctance to modernize its freight cars. Refrigerated cars, cooled by ice blocks, had been available to shippers since the 1840s, but the cost of their construction, operation, and maintenance made them prohibitively expensive for all but the meatpacking industry, which was heavily dependent upon interstate commerce. Indeed, the American canning industry had a vested interest in discouraging the use of refrigerator cars that would have effectively put canners out of business. It is thus hardly surprising that, by the beginning of World War I, America's refrigerated rail cars routinely sat idle during cold months in which frigid temperatures provided boxcars with natural refrigeration. During the war, however, logistical challenges compelled the federal government to force the rail industry to adopt a "car pool" system, under which rail cars were moved to seasonal staging areas to ship critically needed perishables to the military. This resource allocation system endured after the war, and the industrial use of refrigerated cars consequently became much more commonplace. The resulting blow to canneries was compounded by a paradigm shift in postwar refrigeration technology. The introduction of chlorofluorocarbon-based refrigeration equipment in the 1930s revolutionized the food preservation industry by making both industrial and domestic refrigeration truly affordable for the first time.

In South Louisiana, the advent of affordable refrigeration technology coincided with the convergence of several significant historical developments: establishment of the so-called jumbo shrimp market by Morgan City fishermen in the 1940s and 1950s exponentially increased the national demand for the oversized Gulf crustacean. In addition, various Louisiana entities, working in cooperation with the federal

MORGAN'S LOUISIANA & TEXAS RAILROAD.

U. S. MAIL ROUTE BETWEEN NEW ORLEANS, WESTERN LOUISIANA AND TEXAS.

The ferryboat will leave the foot of St. Ann street, daily at 7½ M., connecting with Passenger Train, leaving Algiers at 8 A. M.

Commutation tickets for 500 miles at fifteen dollars, and tickets for all stations on the road sold at office of the undersigned.

All freight must be prepaid.

Freight for the road, for landings on the Teche and for Opelousas, delivered at New Iberia or St. Martinsville; also freight for all landings on Bayou Vermilion, received daily (Sundays excepted) from 8 A. M. to 5 P. M. at the foot of St. Ann street.

Freights for all points on Bayou Lafourche, between Labadieville and Lockport, will be transported at through steamboat rates.

Freight via Houma Branch will be received for and from stations and platforms between Terrebonne and Houma, for and from all points on Bayou Terrebonne between Houma and the plantation of J. B. Bisland, Esq., for and from all points on Little Caillou, between J. B. Bisland's plantation and that of F. J. Pavy, Esq., and from all points on Grand Caillou between P. J. Pavy's plantation and that of Butler.

Say via Houma Branch, and mark destination plainly on all packages.

To insure dispatch freight must be delivered before 8 P. M.; all freight received after that hour will go forward on next day's train.

GEORGE PANDELEY, Superintendent.

For freight or passage, apply at the corner of Natchez and Magazine streets

No freight received after 5 o'clock.

CHAS. A. WHITNEY & CO., Agents,
New Orleans, La.

RENE MACREADY, Agent,
Morgan City.

Daily railroad administration was provided to numerous landings along the right-of-way between New Orleans and Morgan City. Morgan City's rail line provided fast and reliable transportation of perishable products to expanding urban centers east and west of the city. (From the authors' collections)

government, launched a concerted effort to bring rural electrification programs to the coastal plain. Beginning in the late 1930s, these programs, despite interruption during World War II, brought electricity to most remote communities by the early 1950s. Coastal residents rushed to embrace electrification and the improved quality of life it afforded, and by 1960 the area, once a technological backwater, was ahead of the national curve in the acquisition of refrigerators.

Electrification coincided with the extension of the state highway system into formerly physically isolated communities. Although these roadways—built at great public expense—were intended to provide logistical support to the oil and gas industry,

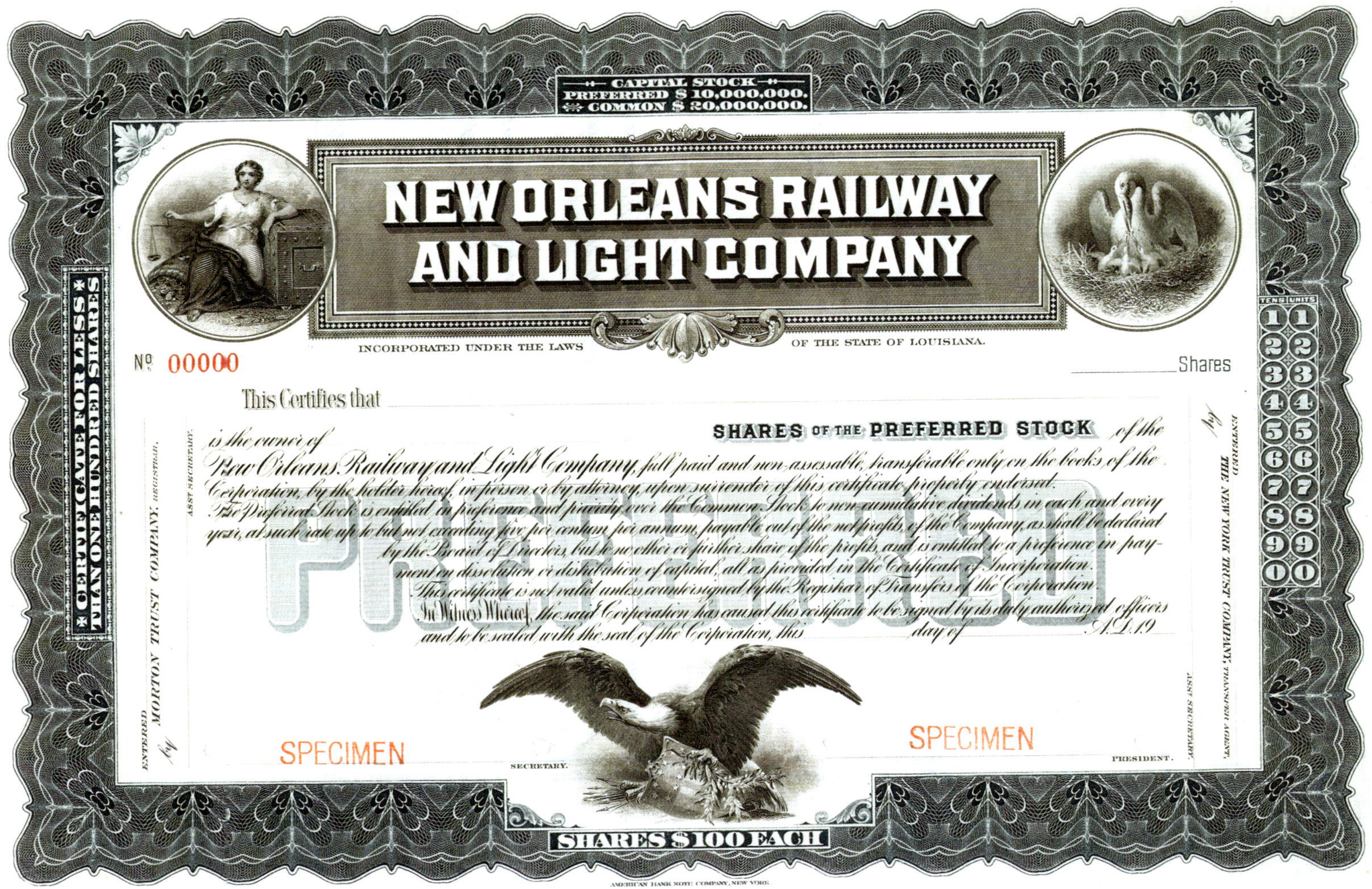

By the beginning of the twentieth century, numerous independent streetcar lines merged to form the New Orleans Railway and Light Company. This company also installed electric lighting along city streets and operated an electric light and gas plant. The company eventually morphed into Orleans Public Administration before absorbed by Entergy. (From the authors' collections)

which was concentrating its exploration and development activity in the coastal marshes and near offshore waters in the post–World War II era, they also inadvertently boosted established economic activities in the newly connected communities. This is particularly true of the seafood industry.

Seafood consumers have always placed a premium on freshness, and, for mid-twentieth-century suppliers, quick, direct access to markets spelled the difference between profitability and ruin. Prior to the coastal road construction initiative, shrimpers, oystermen, and crabbers had to ship their product to processors by boat.

Now, refrigerated trucks could transport their catch from the docks directly to major commercial consumers—grocery stores, seafood markets, fine restaurants, and seafood raw bars.

Direct access to consumers not only broadened the potential market for local producers and enhanced the industry's profitability by cutting out middlemen; it also allowed the Louisiana seafood industry to respond more rapidly and directly to changes in the national marketplace. This is seen perhaps most clearly in the meteoric rise of the Louisiana crab industry in the mid-twentieth century.

The rise of the Louisiana crab industry—like the state's oyster industry—was tied directly to the decline of its counterpart in Chesapeake Bay. In 1875 North America's commercial crab trade originated with the shipment of soft-shell crabs from Maryland. Later, crab cakes became a popular, eventually traditional, dietary specialty and the icon of Chesapeake Bay's crab fishery. A commercial crab fishery emerged to satisfy the growing and seemingly insatiable national demand for both specialized seafood products.

The Maryland experience stands in stark contrast to that of Louisiana, where crabs were simply a modest part of the coast's subsistence economy. In the nineteenth century, Louisiana residents caught blue crabs strictly for home consumption, utilizing trotlines, bait, and small boats.

In Louisiana, the commercial crab harvest began only after ice became widely available to the regional seafood industry. By the 1880s New Orleans hotels, restaurants, and steamboats had become important blue-crab consumers. In addition, crabs became plentiful in the French Market, as marsh dwellers added another income stream to their seasonal harvesting activities. However, without refrigeration and a well-developed processing and transportation infrastructure, the crab market was severely limited geographically. In addition, the catch was valued at about a one cent a dozen, a price that was uncompetitive nationally. As a result, although Louisiana supplied the local market with more than 1.4 million pounds of crabmeat in the late 1890s, according to the US Commission of Fish and Fisheries, its crab fishery for decades remained a regional player in a national market dominated by more established Chesapeake Bay crabbers.

In 1908 Louisiana ranked third nationally in crab production, and Jefferson, Lafourche, Orleans, and St. Bernard Parishes provided most of the crabs shipped to retail and wholesale outlets in New Orleans—the largest seafood market on the Gulf Coast. The canning industry did not become a viable outlet for crab fishermen until the second decade of the twentieth century, when a processing plant was built

Large-scale canning operations made seafood products available for the first time to national markets. (Photo from E. P. Churchill, *The Oyster and the Oyster Industry of the Atlantic and Gulf Coasts*, Appendix VIII, Report, 1919, US Bureau of Fisheries, Doc. No. 890 [Washington, DC: US Government Printing Office, 1920])

The Dunbar name has been associated with the canning industry since 1865, when George H. Dunbar began canning semitropical products, such as fresh shrimp, potted shrimp, green turtles, preserved figs, orange preserves, figs in cordials, okra, and other fruits and vegetables. The company's international markets included England, France, Germany, and Spain. (Photo by the authors, 2011)

For decades, Louisiana's rice fields served as the principal source of frog legs for America's premier restaurants. (Photo courtesy of the National Archives and Records Administration, Prints & Photographs Division, College Park, MD, photographer Russell Lee, 1938, call number 8a2433a)

in Morgan City. In the 1930s Louisiana's Department of Conservation reported the birth of the state's crab canning industry, but failed to provide statistics regarding the regional crab harvest. The production side of this rapidly evolving business was evidently either inadvertently underreported or considered too insignificant to be included in the official reports. The department nevertheless stressed the state's need to further develop the nascent crab industry. The agency urged fishermen to abandon inefficient trotlines for the "crab pot" technology used by Chesapeake Bay watermen.

These crab traps, consisting of a wire frame box designed to lure the crab into the enclosed chamber, permitted fishermen to increase their catch dramatically. The timing of this paradigm shift in the Louisiana crab fishery's harvesting technology was most fortuitous, for it coincided with the decline of the then nationally dominant Chesapeake Bay crab fishery. Louisiana, where blue-point crabs are extremely abundant, was in the perfect position to fill the void. Beginning in the 1970s the state's crab fishermen shipped live crabs by truck—later by airfreight—to eastern markets. This practice continues, with prices in 2014 as high as $3 a pound and number one Louisiana crabs reportedly selling in Baltimore for $100 per dozen—a far cry from the one cent a dozen the state's fishermen received at the dawn of the twentieth century. The financial return is so great that many large coastal landowners allow commercial crabbers to harvest these crustaceans on their properties through lease agreements.

Because demand perennially exceeds production—particularly internationally as a result of rising Asian standards of living—crabs and crabmeat have evolved into luxury food items, and Louisiana's coastal parishes constitute the backbone of the national crab fishery. Although Chesapeake Bay crabbers established the American crab fishery, Marylanders now regularly—though unknowingly—consume Louisiana crabs. Louisiana's crab industry presently employs, directly or indirectly, about 3,000 individuals who, since 1997, have annually harvested more than forty million pounds with a market value of approximately $30 million.

Louisiana's recreational crab harvest adds significantly to the commercial catch statistics, but this catch is sadly undocumented. A weekend journey into the crab's primary habitat nevertheless clearly reveals that this is a tremendously popular family practice.

Like the crab industry, Louisiana's frog export business rose to national prominence in the early twentieth century, but, unlike its sister industry, it lacked the staying power to remain an important regional economic driver. Although thirty frog species are native to Louisiana, the American bullfrog had the greatest commercial value because of its role as an exotic staple of late-nineteenth and early twentieth-century

Franco American haute cuisine. Louisiana's participation in the frog leg trade evidently began in the late 1880s, when French immigrant Donat Pucheu settled at Rayne, a newly established rail town in Acadia Parish.

Pucheu, an enterprising individual, quickly profited from the convergence of fortuitous circumstances in his adopted hometown. The coming of the railroad to Louisiana's central prairie region in 1880 had revolutionized the area economically, and within a decade the area had become one of New Orleans's principal suppliers of eggs and other agricultural commodities. Prominent among these agricultural products was rice, which, thanks to new German and midwestern immigrants in the Rayne, Crowley, and Iota areas, had quickly emerged as the dominant local crop. Rice paddies produced ideal breeding grounds for bullfrogs (known locally as *ouaouarons*).

Pucheu was well aware of *ouaouarons*' commercial value to French restaurants, particularly those in New Orleans and New York, which had been the nation's leading ports of entry for Gallic immigrants in the nineteenth century. Pucheu thus had an abundant supply of frogs; easily accessible rail links to New Orleans, which the region was already utilizing to export large quantities of eggs and rice; and a ready availability of chronically cash-strapped local farmers, tenants, and sharecroppers always eager to find means of supplementing their income—particularly during the spring and early summer months, when the rice fields were flooded.

Other local entrepreneurs eventually supplanted Pucheu as the local bullfrog processor and exporter. Jacques Weill, who emigrated from Paris in 1901, joined local businessmen to establish the firm of Jacques Weill, Boudreaux, and Leger, specifically for the purpose of shipping dressed local frogs to external markets, which by the 1930s included Chicago, Cleveland, Cincinnati, Pittsburgh, and Los Angeles.

Jacques Weill's success spawned a multitude of local and regional competitors, resulting in the South Louisiana frog industry's eventual expansion into other communities, including St. Martinville, Pierre Part, and Des Allemands. At the industry's peak, the area's frog dealers dressed and shipped approximately 400,000 frogs (about 2,000,000 pounds of meat) annually. Various factors, however, caused the regional output to decline steadily during and after World War II.

Manpower shortages during the war and the rapid disappearance of tenants, sharecroppers, and small farmers after the war, as well as the catastrophic impact of DDT upon the regional amphibian population during the postwar years, had devastating cumulative consequences. By 1949 Louisiana's exports had declined to only 200,000 frogs, and by 1970 the industry had largely disappeared, and Mexican,

More than 100 years after the Pennsylvania industry's decline, the fracking industry in the Marcellus formation has resulted in the local hydrocarbon industry's rebirth. In 2015 the Keystone State was the United States' second largest producer of natural gas, supplanting Louisiana.

Japanese, and Indian exporters had supplanted Louisiana dealers and the major national frog leg suppliers.

The demise of Louisiana's frog leg industry coincided with the rise of the region's crawfish industry, which conveniently filled the economic void, causing, in the process, a culinary revolution in the coastal plain. Crawfish, like lobsters in the Northeast, were traditionally a "poor man's food." In predominantly Catholic South Louisiana, residents historically consumed them primarily during Lent, when religious dietary proscriptions banned the consumption of meat. Crawfish provided an abundant and convenient protein alternative, but because they were typically boiled without seasoning, crawfish had little popular appeal to early twentieth-century local consumers. In fact, after the 1927 flood, American Red Cross workers lamented the fact that refugees did not readily consume the crustaceans outside the Lenten season. The following decade, Percy Viosca, described by some as Louisiana's homegrown Audubon, also recognized the value of the neglected resources and recommended seasonal crawfish harvesting by means of traps in managed ponds. Viosca's appeal went unheeded, particularly among rice growers whose rice paddies constituted ideal crawfish habitat in prairie parishes. The farmers' persistent apathy would evaporate only after 1958, when four developments would radically alter the embryonic industry's ostensibly gloomy prospects.

The industry's success hinged upon its ability to establish and nourish a market for commercially harvestable native crawfish. Approximately 100 crawfish species exist in North America, but only two indigenous varieties are available in sufficient abundance and size to warrant commercial harvesting. These species are popularly labeled "pond" and "river/swamp" types. Commercial crawfishermen had traditionally harvested the river/swamp variety, particularly in the Atchafalaya Basin, and much of this catch was sent to upscale New Orleans restaurants, where it was transformed into crawfish bisque—the Creole dish par excellence.

As Percy Viosca's quixotic campaign had demonstrated, however, mere raw product availability did not a market make. In 1959 industry boosters in Breaux Bridge established the popular Crawfish Festival to promote the product. Then, in the late 1960s, packaged, peeled crawfish tails became readily available, and restaurants subsequently began to feature the crustacean on their menus—usually in the form of crawfish étouffée, which became a regional culinary sensation. In addition, as a result of steadily rising local family incomes during the 1960s oil boom and the emergence of two-family incomes in the 1970s, which greatly expanded the amount of disposable income in typical family budgets, regional consumers had the wherewithal to buy such processed "luxuries." Finally, women in two-income households quickly learned to appreciate the convenience of packaged crawfish tails. Crawfish bisque typically required two days of preparation in the early-to-mid-twentieth century; however, a cook using processed tails could prepare crawfish étouffée in as little as thirty to forty-five minutes.

The crawfish industry deftly adapted to meet the skyrocketing regional demand. Because of the small number of crawfishermen in the Atchafalaya Basin as well as the crawfish "crop's" heavy dependence on water levels that fluctuated dramatically seasonally, another, more reliable source of supply was essential. Rice farms promptly filled the void, as farmers, who had recently lost the vital Cuba market as a result of the postrevolutionary embargo, desperately sought to mitigate the resulting economic fallout.

South Louisiana rice paddies naturally teem with crawfish, and rice farmers discovered they could turn the crustaceans, which often burrowed into impoundment levees, from a nuisance to a harvestable resource essential to their annual bottom line. South Louisiana's rice farmers, who had long experience with irrigation systems and impoundments, quickly realized that participation in the emerging regional crawfish industry posed few technical challenges, required minimal expenditures, and offered an ideal solution to their crop rotation quandary. Crop rotation had previously forced farmers to either allow the paddies to lie fallow for a season (with the attendant loss of income) or endure the expense and aggravation of dismantling the levees to permit cultivation of row crops. Crawfish "farming" entailed no significant technological or mechanical investment, and it permitted farmers to earn significant income from their "fallow" land, which was naturally fertilized by water-transported sediments, crawfish shells, and, in more recent years, vegetation planted to nourish the omnivorous crustaceans.

Virtually overnight, crawfishing became a major South Louisiana industry. Although the number and size of ponds vary annually, the amount of land devoted to farm-raised crawfish has occasionally reached nearly 120,000 acres—nearly 190 square miles. The regional market now depends on the perennial reliability of the pond-based fishery.

As the number of crawfish ponds expanded, so did the emerging industry's attendant infrastructure. Entrepreneurs established new processing plants to service the market. Supplanting the modest, home-based family businesses associated with the industry's infancy, these new facilities, housed in easily recognized metal buildings, can handle several thousand pounds of crawfish per day. These large operations afford consumers not only the economies of scale but also the high qualitative standards demanded by the marketplace. Nine supervised steps are required to transform raw crawfish (delivered in polypropylene "onion" sacks) into a peeled, cooked, inspected, and vacuum-packed product ready for distribution to restaurants and supermarkets.

Crawfish processing has traditionally been a labor-intensive industry, despite the availability of mechanized peeling devices, which have been in existence since Fernand S. Lapeyre of Houma first patented a machine in 1961. Many discerning Louisiana consumers are happy to pay premium prices for handpicked crawfish meat—which typically contains fewer shell fragments per pound. Many processors consequently still rely upon seasonal laborers to peel, clean, and inspect the catch—these are H2B seasonal immigrant guest workers who provide most of the seafood processing plants labor.

The success of South Louisiana's crawfish industry has overshadowed and obscured other regional experiments in aquaculture. Turtle harvesting is the oldest of these competing industries. Indigenous turtle varieties evidently had little market value in the Bayou State until the turn of the twentieth century, when snapping and alligator snapping (loggerhead) turtles began to appear in local markets. Although modest numbers of these turtles were eventually harvested in the Atchafalaya Basin and other South Louisiana swamps for area restaurants producing turtle soup in the early twentieth century, Prohibition and the disappearance of sherry—considered by connoisseurs as an essential complement to turtle soup—effectively killed the industry by 1933.

The demise of the snapping turtle business contributed to the emergence and growth of the green turtle industry. By the end of Prohibition in 1933, swamp dwellers had begun to raise red-eared sliders or green turtles in impoundments constructed expressly for this purpose. These "farms," most of which were located in the Pierre Part area, exported turtles to China, where they were used for food and in traditional medicine, and to American cities to satisfy the turtle pet market that exploded in the wake of the Chicago World's Fair of 1933. However, the discovery that turtle shells could be contaminated with salmonella—bacteria causing fever, stomach cramps, diarrhea, nausea, and, in extreme cases, death—eventually led to the industry's demise. Between 1933 and the early 1970s, more than 100,000 Americans would be diagnosed with turtle-transmitted salmonella. Responding to this perennial health threat, the federal Food and Drug Administration (FDA) banned pond-raised turtles from the pet market in 1975.

The FDA ban forced the industry to adapt. Research funded in large part by Louisiana's Sea Grant program resulted in the development of the Siebeling method (devised by Ron Siebeling) of sanitizing eggs to produce nearly salmonella-free turtles. Once eggs hatch, Louisiana's state-approved laboratories test the hatchlings, certify those found to be salmonella free, and secure governmental certificates authorizing distribution of the chelonians.

The new certification system has permitted the turtle hatchery industry to initiate a rebound throughout the Deltaic Plain. Because the FDA bans the interstate transportation of turtles with shells less than four inches in diameter, Louisiana turtle farmers have begun to export the chelonians to international markets. The export industry nevertheless remains small; in 2014 sixty turtle farmers produced more than ten million hatchlings with a market value of nearly $5 million.

As the turtle industry's experience makes abundantly clear, wholesale regulatory and environmental changes have radically altered the business climate and long-term prospects for many, if not most, operations commercially exploiting aboveground resources. The challenges are perhaps even greater for companies engaged in subsurface extractive industries. Development of coastal Louisiana's below-ground mineral resources has traditionally focused upon salt, sulphur, and hydrocarbons, which often exist in close proximity to the salt dome formations that riddle the coastal parishes. Salt domes—with elevations of up to 230 feet—are the region's most imposing geological features, and because salt is easily the most accessible of the three minerals associated with salt domes, it was the first to be exploited commercially.

Salt domes—or diapirs—are colossal vertical shafts of salt, covered by mantles of caprock, consisting in part of sulphur. Were these diapirs aboveground, they would tower over Mount Everest. The largest Louisiana salt domes have tops (actually round hills) measuring from six acres to twelve square miles. There are nearly 500 domes along the upper Gulf Coast, but the most economically significant of these are all located in the Louisiana and Texas coastal plain. While the most physically imposing of these geological structures once boasted notable agricultural operations dating to antebellum times, they became much more noteworthy in the twentieth century for their strategically important mineral deposits.

The largest visible Louisiana salt domes—Belle Isle, Weeks Island, Avery Island, Jefferson Island, and Cote Blanche—are located along a southeast–northwest axis on the central Louisiana coast. In 1862 workers excavating a brine spring discovered rock salt at Avery Island. This mine constituted the Confederacy's principal source of salt—an essential food preservative in the pre-refrigeration era—until invading Federal forces destroyed the facility in 1863. During its brief lifetime, the Avery Island mine generated approximately 20,000 tons of salt. Salt operations resumed shortly after the war's conclusion, and the Avery Island mine is now the nation's oldest in terms of continuous subsurface operations—perhaps the oldest in the Western Hemisphere. Sinking shafts to approximately 1,000 feet and mining salt by the "room-and-pillar" method, the Avery Island salt operation created rooms measuring 70 to 90 feet in height, supported by columns measuring approximately 70 by 70 feet. Machines used to excavate rock salt are disassembled, transported into the mine by elevator, and then reassembled.

Because of Avery Island's economic importance, state mining officials routinely documented the mine's output in their annual departmental reports. These documents indicate that, in the late nineteenth century, local salt output fluctuated annually between 25,000 and 45,000 tons—moving in lockstep with the vagaries of the state and national economies. Salt mining operations began on neighboring Weeks Island in 1902, and the Louisiana salt industry quickly developed a price advantage over its northern rivals, particularly Michigan salt operations that extracted the mineral from either "The Great Salt City" 1,200 feet beneath Detroit or saline ponds and springs through expensive evaporation processes. Between 1900 and 1904, for example, Michigan's output of coarse salt declined by 26.9 percent, while the reported output of Louisiana mines rose by a staggering 474.4 percent. Over the following two decades, the output of Louisiana's salt dome mines consistently ranged between 150,000 and 190,000 tons. Output would plateau at 300,000–500,000 tons in the 1920s, when commercial salt production began at Anse La Butte (St. Martin Parish) and Jefferson Island (Iberia Parish), before achieving new heights—over 775,000 tons—in the mid-1930s as a result of salt extraction from brine at Hackberry (Cameron Parish).

Over the past century, Louisiana salt has been used primarily in industrial applications. Salt mined at Avery Island and other South Louisiana salt domes, for example, is annually barged to northern states for use on highways and driveways in winter months. In fact, Louisiana is currently America's leading salt producer, generating almost one-fourth of the nation's output.

Coastal domes, which collectively remain Louisiana's principal industrial salt source, are sometimes overlain by a caprock that contains significant amounts of elemental sulphur. Some salt domes contained sulphur lenses large enough to warrant commercial extraction. This was the case in coastal Louisiana, where the mineral was recovered by drilling a well into the caprock, pumping superheated water (165° C) and compressed air into the formation, respectively, to dissolve the mineral and force it to the surface by the down-hole pressure.

This mining process, perfected by the chemist Herman Frasch at Sulphur, Louisiana, in 1903, typically extracts from 75 to 90 percent of a salt dome's sulphur deposit at 99.5 percent purity. Mining companies soon successfully utilized Frasch's technology at other coastal salt domes—including the Port Sulphur area of Plaquemines Parish—and, as a result, Louisiana quickly emerged as the global bellwether in sulphur production.

At the industry's peak in the 1980s and very early 1990s, Louisiana boasted thirteen land-based sulphur mines and one offshore facility. By the mid-1990s, however, Louisiana's established sulphur producers faced insuperable competition from the state's homegrown petroleum industry. Most sulphur is currently produced as a by-product of crude oil refining and natural gas processing. Sulphur extraction from salt domes was no longer cost competitive, and the industry experienced a rapid decline and untimely demise. The Main Pass 299 mine in offshore waters, Louisiana's last Frasch-method sulphur producer, suspended operations in late 2000.

Like its erstwhile competitor, Louisiana's early petroleum industry was closely linked to mineral extraction at salt dome sites. The state's early oil exploration resulted directly from the growing national demand for petroleum products, particularly kerosene. Prior to the 1850s, many Americans used whale oil in household lamps. The oil was a cheap illuminant, but it produced a pungent smell that, for many consumers, negated its commercial advantages. In addition, whale oil, especially the superior sperm oil, became increasingly scarce—and hence more expensive—due to overfishing in the decade before the Civil War.

A fortuitous convergence of events allowed the embryonic petroleum industry to eclipse the whale oil industry during the Civil War, when the threat of Confederate privateers forced many New England whalers to remain in port. Canadian Abraham Pineo Gesner's invention of coal-derived kerosene in 1846 was subsequently complemented by the Polish chemist Ignacy Łukasiewicz's

In the early 1960s gasoline was thirty-two cents a gallon, and like these rusting vintage oil pumps, that price is history (above). In addition, in the coastal plain, and elsewhere, kerosene lamps were the most common source of domestic lighting until rural electrification (left). (Photos by the authors, 2011)

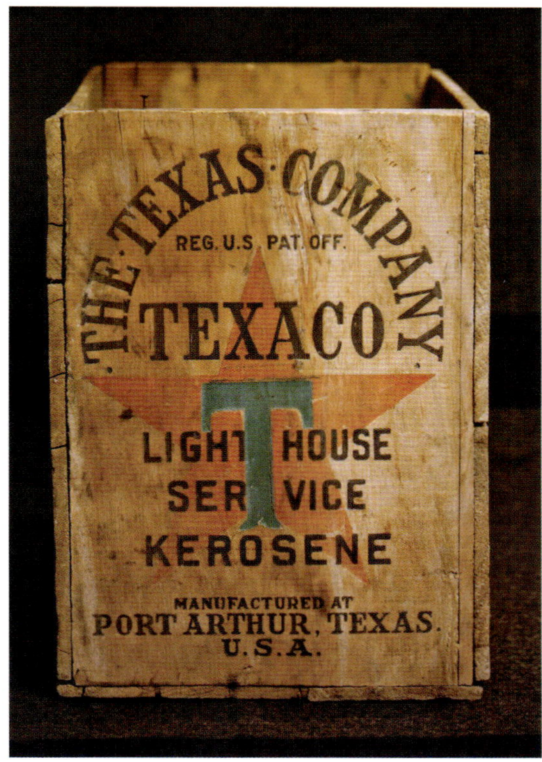

Manufactured in Port Arthur, Texas, with Louisiana wood, the Texas Company's cypress cases were a mainstay of the American kerosene export trade. Kerosene packed in five-gallon tins (two tins per case) was a staple of America's early twentieth-century export trade. Petroleum companies, on the other hand, shipped crude oil to external markets in barrels. (Photo courtesy of Steve Gronow of the Maritime Exchange Museum, Howell, MI)

Extraction of sulphur by the Frasch process was a precursor to the Louisiana oil and gas industry. (Photo courtesy of Fred Dunham, date unknown)

The American sulphur industry was born in Louisiana in the last decade of the nineteenth century. (Photo by the authors, 2011)

discovery of a method of distilling kerosene from petroleum and his invention of the kerosene lamp. Development of the Titusville, Pennsylvania, oil field in 1859 and the oil boom that it created tipped the scales in favor of the Łukasiewicz method. By the 1870s the Keystone State was a national leader in petroleum production, but Pennsylvania's output peaked around 1891 before beginning a protracted decline.

Pennsylvania's declining output gave impetus to oil exploration elsewhere in the United States, and the resultant 1901 oil strike at Spindletop, near Beaumont, Texas, immediately shifted the petroleum industry's focus to the Gulf Coast. Oil fever quickly spread to Texas's eastern neighbor, and it reached a fever pitch when W. Scott Heywood brought in a gusher near present-day Evangeline (near Jennings, Jefferson Davis Parish) in September 1901. With development of the hugely successful Evangeline field, Louisiana assumed a central role in the nation's industrialization, which was increasingly powered by oil and gas.

After 1901 South Louisiana communities treated itinerant wildcatters like visiting royalty in attempts to entice them to remain long enough to examine the local geology for potential plays. As a result, major new fields were brought into production in St. Martin Parish and other coastal plain communities by 1905. These new fields consisted of dozens of closely clustered shallow wells. In fact, by 1906 the Evangeline field boasted a veritable forest of wooden derricks in such close proximity that one could literally step from one derrick floor to another without touching the ground.

The explosion in oil production necessitated the equally expeditious development of storage and transportation infrastructure. Nearly every well brought into production during the first decade of the twentieth century generated 2,500 to 17,000 barrels daily. One particularly prolific well produced 1.5 million barrels in 145 days. (With a "head" diameter of twenty inches, these barrels, when placed side by side, would have stretched nearly 5,700 miles.) The Evangeline field's production was particularly noteworthy. In 1904 the play's output approximately equaled the peak performance of the fabled Spindletop field.

To handle some of the output of the Evangeline field, oilmen established a barge-loading facility along nearby Bayou Nezpiqué. Movement of petroleum in bulk by barge was problematic in the early twentieth century. Many inland fields were distant from navigable streams, and even when oil fields could be serviced by waterborne transportation, barges faced difficulties in reaching major ports and, later, refining facilities. Before completion of the Gulf Intracoastal Waterway in 1942, major east–west internal waterways did not exist along the upper Gulf Coast. Hence transporters had to tow barges along the coast, where they were highly susceptible to storms and

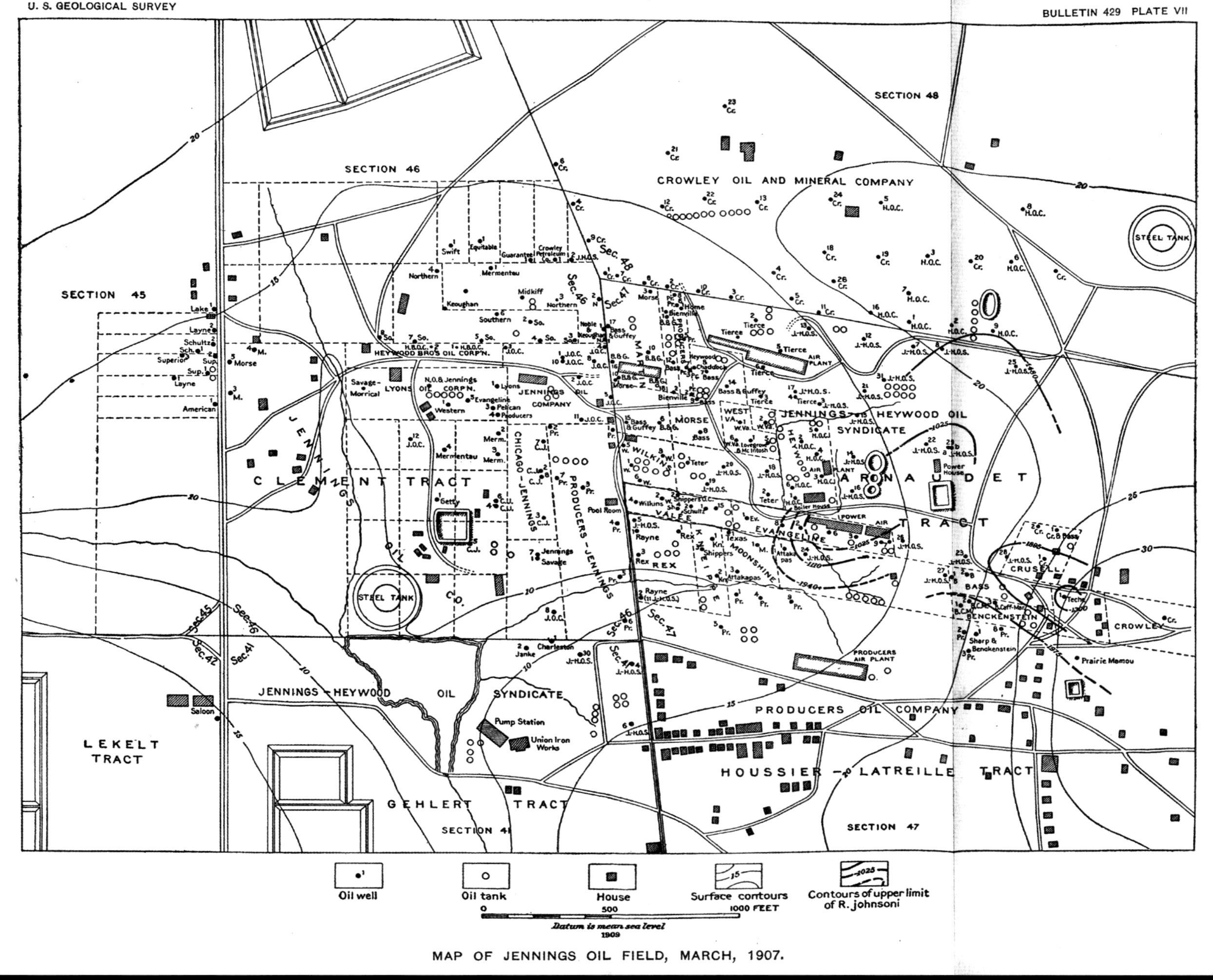

Nine years after the discovery of oil at Jennings (actually Evangeline), the Clement Tract, the site of the first well, was covered with wells, storage facilities, outbuildings, and a plethora of small independent oil companies. (Map from G. D. Harris, "Oil and Gas in Louisiana with a Brief Summary of Their Occurrence in Adjacent States," *U.S. Geological Survey Bulletin*, 429 [1910] Plate VII)

Oil and gas exploration and development in southern Louisiana initiated a bitter culture war that has endured to the present. In the early twentieth century, wildcatters and their drilling crews were almost exclusively Texan. These brash, often arrogant interlopers clashed immediately with South Louisiana's established French-speaking population, whose ethos was framed by laissez-faire egalitarian individualism. These frequently acrimonious skirmishes—waged across a variety of fronts, including Sunday blue laws, quixotic local option prohibition initiatives, and efforts by unscrupulous oilmen to fleece mineral rights from illiterate Francophone Cajuns, Creoles, and Native Americans—resulted directly from largely irreconcilable worldviews. The regional culture war combatants' worldviews were framed by religion, language, class, and Progressive ideology. The transplants, steeped in Bible Belt evangelism and Progressivism's validation of America's mainstream Protestant values, felt no compunction about attempting to impose their values on the new oil province's huge Catholic majority.

This cultural onslaught occurred at precisely the time when Louisiana public schools—as agents of American Progressivism's cultural homogenization crusade—sought to eradicate the indigenous population's French language and culture, and the regional English-language print and electronic media openly lampooned Francophones. As one might expect, this did little to endear the immigrants to the indigenous population.

The gulf dividing the Texas and Louisiana populations was widened still further by the glass ceiling faced by Catholic workers—an impervious barrier that survived in some oil companies until the post–World War II era. (Cajuns, for example, usually could not rise above the grade of driller in many oil companies, while Anglo Protestants [most of them Masons], on the other hand, monopolized the administrative positions.)

The cultural contest abated after Pearl Harbor in the interest of national unity during World War II, only to resume after VJ Day. As testimony in the *Roach v. Dresser* (494 F.Supp. 215, D.C. La., 1980) case clearly indicates, the cultural war was alive and well nearly three decades later, even as the glass ceiling disappeared to accommodate college-educated, upwardly mobile Cajun and Creole Baby Boomers. The old wounds continue to fester in many coastal communities, where out-of-state oil workers are *still* known as *les maudits Texiens* (the damned Texans).

Oil-Mad Throng in Rush to Louisiana

Three Wells Yield 50,000 Barrels Oil

Nov. 12---15,000-Barrel Well
Nov. 12---20,000-Barrel Well
Nov. 12---15,000-Barrel Well

All Records Smashed by Production of Oil in Claiborne Field. Three Wells in One Day, and All Gigantic Gushers, Breaks the Record.

Nov. 22, 20,000-barrel well. Nov. 22, another 20,000-barrel well.

The Unprecedented Production of Oil in North Louisiana Is the Marvel of the Age

The Wonder Gushers of the World Brought In in the Famous Claiborne and Caddo Oil Fields

Oct. 27 a 35,000-Barrel Oil Well. Largest in the World—High-Grade Oil

Oil is King in the World Today

Financial interests the world over recognize Caddo and Claiborne Oil Fields as the greatest in the World. We OWN lands and leases in these great fields.

Some of our land is surrounded by wells. We also own leases in Texas and Terrebonne Parish, La. Terrebonne has the largest gas well in the world.

The Fawn Creek Oil & Gas Co., have 893 acres in these great fields.

Fawn Creek Stock Now $1.00 Per Share Par Value

We are offering a small block of stock—NOW—at $1 per share. The stock is due for an IMMEDIATE advance. It is, we believe, your chance of a LIFETIME to make BIG MONEY out of a small investment.

Active Drilling Operations Will Start Shortly. Our Driller Is Now in the Field.

OFFICERS AND DIRECTORS

M. J. SAMUELS, New Orleans, La., President. Member of Firm of Samuels & Co., Merchant Tailors, Chicago, Memphis, New Orleans, Davenport, Peoria, Etc.

A. P. MORESI, Jeanerette, La., Vice-President. Prominent Merchant. Owner of large Machine Shops and Foundry, Ice Plant and Bottling Works.

F. C. DENGLER, New Orleans, Secretary. Is an Expert Accountant and Stands High in the Community.

F. E. BAUDRY, Convent, La., Treasurer. Prominent Merchant and Planter.

HON. N. T. BOURG, Thibodaux, La., Director. Mayor and Prominent Real Estate Owner and Sugar Planter.

F. B. GREVEMBERG, New Orleans, Director. Manufacturer's Sales Representative and Real Estate Owner.

DR. B. B. WARREN, Covington, La., Director. Prominent Physician and Surgeon for I. C. Railroad.

Write at Once for Full Particulars and Prospectus. It Don't Cost You One Cent to Investigate.

FAWN CREEK OIL & GAS CO.

611 CANAL BANK BUILDING, NEW ORLEANS, LA.

Louisiana's early twentieth-century oil "rush" transformed the coastal plain's economic landscape. Pioneer drillers often issued stock to underwrite their exploration activities. (*New Orleans Herald*, December 18, 1919)

The Louisiana oil and gas industry was born at the Evangeline field, near Jennings. Fire was a constant threat at early production sites. (Photo from *The Jennings Gusher Fire*, 1902, Rein Litho. Co., courtesy of the Beinecke Library, Yale University, call number Zc22+902je)

heavy seas. Shippers consequently faced onerous insurance charges in addition to crushing transportation costs.

These cost factors, coupled with falling petroleum prices between 1904 and 1905 and the relative difficulty of drilling many of the most promising stratigraphic traps, particularly in the coastal plain, persuaded many wildcatters to relocate to North Louisiana.

Between 1904 and 1906, exploration and development was heavily concentrated in the Caddo Lake area, about fifty miles northwest of Shreveport. By 1910 three sides of the lake were lined with nearly 200 producing wells. The Caddo Lake field enjoyed a dramatic second growth spurt after 1911, when the Gulf Refining Company (born at Spindletop in 1901) successfully completed the world's first oil well over a natural water body. This well, which also enjoyed the distinction of being the nation's first successful maritime-drilling operation, resulted in a second wave of intensive exploration and development. The high success rate at Caddo Lake provided the impetus for sustained, concentrated exploration and production in Caddo and DeSoto Parishes, and later in Ouachita Parish.

These North Louisiana experiences provided the petroleum industry with invaluable training for the next chapter in its regional development. This new historical phase was driven by the post–World War I emergence of a national highway system and the transportation revolution that it engendered. Between 1910 and 1920, auto and truck sales surged from less than 500,000 to more than nine million, and the national market demand for fuel grew accordingly. The swelling national demand for petroleum products—particularly gasoline—compelled the industry to revisit promising South Louisiana sites in the 1930s.

Along the Louisiana coast, expanding production meant exploration and development operations in environments previously shunned by wildcatters. This meant utilization of over-water drilling technologies developed earlier at Caddo Lake. It also meant transplantation of transmission technologies also perfected in North Louisiana oil fields. The enormous productivity of the interior fields—Caddo, DeSoto, and even Oklahoma's Glenn Pool (once the world's largest)—easily outstripped the industry's ability to export the raw product from the oil fields.

The American oil industry was consequently forced to confront the twin challenges of transporting and refining unprecedented quantities of petroleum. The first near coastal fields had permitted the use of cheap waterborne transportation, but movement of oil from landlocked interior fields by conventional overland means was not an option in backwater regions, in which numerous important early oil provinces

For over the past half-century, the near- and offshore oil and gas industry has been a mainstay of coastal Louisiana's industrial economy. (Photo by the authors, 2003)

The Glenn Pool oil field, discovered in 1905, caused an economic boom that propelled Tulsa into economic prominence. The community now bills itself as "The Town That Made Tulsa Famous." This slogan is equally true for Baton Rouge, for Glenn Pool oil, transported through nearly 10,000 miles of Standard Oil pipelines, was instrumental in the construction of Standard Oil's Baton Rouge refinery in 1909. (Photo by the authors, 2010)

The early Vinton (Ged) oil field was a veritable forest of wooden oil platforms. (From the authors' collections, used with permission of David Richard)

Loading gasoline for delivery in the coastal zone. Refueling operations required steel barrels, a barge, and a stout push-boat. (Photo from the J. D. Theriot collection)

developed. In many of these areas, adequate overland transportation infrastructure simply did not exist, and, where transportation was available, it was not only unacceptably slow but also prohibitively expensive.

Economic exigencies also forced the oil industry to establish major refineries along the Mississippi River between Baton Rouge and New Orleans, where oceangoing tankers could efficiently export gasoline to East Coast and international markets. The Standard Oil facility at Baton Rouge—America's second oldest in terms of continuous operation—is the first and most notable of these. These facilities were complemented by carbon black plants and other allied industries, and, of course, the pipelines necessary to sustain them—at an estimated cost of more than $250 million by 1930 ($3.55 billion in 2015).

To move petroleum from the fields to the refineries, oil companies had to develop pipelines, and this technology had been perfected in the North Louisiana oil provinces. Screw pipe connections had been replaced by welded joints, and the pipe itself had been upgraded through the introduction of improved steel. These and myriad other critical technical infrastructure improvements permitted the industry's embryonic regional pipeline system to expand nationally. By the 1920s more than 115,000 miles of pipeline crisscrossed the country. Much of the new construction was concentrated in South Louisiana, where, by 1933, approximately two million acres were under lease for oil and gas exploration. Within five years, major fields were under development at Cameron Meadows, Sweet Lake, Bayou Blue, Welsh, Leeville, Dog Lake, Lake Pelto, Iberia, White Castle, Black Bayou, Caillou Island, Ville Platte, Bosco, Lake Barré, East Hackberry, Tepetate, Jeanerette, and Vinton.

The convergence of expanded field operations and new transportation infrastructure permitted South Louisiana to become the in-state leader in oil and gas production for the first time since 1906. Investment followed production, and by the early 1930s oil and gas companies had authorized expenditures in excess of $250 million in Louisiana's southern parishes. Considerable additional investment would be required as the petroleum industry concentrated ever more of its exploration activities in the region's forbidding swamps and coastal marshes.

Because of the topography and inaccessibility of newly developed wetlands fields, as well as the exorbitant cost of constructing conventional roadways in paludal environments, oil companies were compelled to devise cost-effective means of accessing the most promising drilling sites. In the coastal wetlands, suction or bucket dredges cut access channels through the marshy "wastelands," often removing in the process more than a million cubic feet of organic material per linear mile. Once the canals were completed, barge-mounted rigs were floated to their respective drilling locations.

Dredges eventually linked individual canals into intricate watery networks, as well as sites clustered in remote wetlands fields. These networks, in turn, usually connected indirectly—through natural waterways—with the Gulf Intracoastal Waterway, which permitted the transfer of heavy equipment from fabrication yards to staging areas, and, finally, to drilling sites. This logistical system would permit the petroleum industry to complete more than 67,000 wells in Louisiana's coastal parishes over the course of the twentieth century.

Future exploration and development focused on near-offshore waters. In 1937 the Pure Oil and Superior Oil companies built, on Gulf of Mexico State Lease No. 1, a freestanding structure in 14 feet of water 6,000 feet south of Creole, a seaside community in Cameron Parish. The success of this high risks–high rewards exploratory operation—completed in 1938—marked the beginning of the nation's critically important offshore oil industry.

Nine years later Kerr-McGee, at that time an Oklahoma independent, completed a well 10.5 miles offshore in Ship Shoal Block 32. This pivotal event launched a black-gold rush into the Gulf waters beyond sight of land. In these unprotected waters, success was based on ever-improving technology, facilitated by Shell Oil's decision, in 1963, to share its then cutting-edge deepwater technology, the centerpiece of which was the world's first semisubmersible drilling unit. This historic event marked the beginning of the industry's deepwater drilling programs and ensured Shell's competitors would be able to bid on federal deepwater leases. As a result of offshore

Ged, namesake of John Geddings Gray, was an oil boomtown associated with the 1910 discovery of the Vinton (Ged) oil field. At its zenith, the community reportedly had a population of approximately 3,000. By the mid-1940s most of the town had been dismantled, and little remains of this once thriving community. (From the authors' collections)

Early twentieth-century reclamation entrepreneur Edward Wisner was the father of the Louisiana Land and Exploration Company. Wisner attempted to apply the Dutch land reclamation model to Louisiana's near sea-level marshes, but his efforts were thwarted by the 1915 hurricane, and his resulting financial reverses eventually resulted in foreclosures. Midwestern investors, led by Henry Timken, ultimately acquired Wisner's interests with the intent of leasing the land to fur trappers. In 1927 these investors transferred title to these lands to the newly created Louisiana Land and Exploration Company, which was to manage the landholdings (which had reportedly grown to 600,000 acres) and promote oil exploration on these properties. (From the authors' collections)

exploration and development, Louisiana became a national leader in energy production. From 1950 to 1980, the Pelican State consistently ranked in the top three nationally in energy output, accounting for approximately 30 percent of domestic production.

The movement into ever-deeper offshore drilling sites gave impetus to several notable evolutionary changes within the petroleum industry. First, there was a fundamental change in the composition of the regional workforce. Huge, specialized fabrication yards built along the coast to construct ever-larger offshore drilling platforms attracted large numbers of skilled blue-collar laborers (welders, electricians, pipe fitters, fabricators, and so forth) from the agricultural sector in the coastal plain. In addition, farmers, tenants, and sharecroppers without industrial skills flocked to the offshore rigs as roustabouts, particularly after Hurricane Audrey destroyed the 1957 cotton crop. Second, major corporations began to establish regional offices in New Orleans, Houma, and Lafayette to administer offshore activities. Third, independent service industries began to cluster around the district headquarters specifically to service operations in Gulf waters.

Service companies such as Schlumberger, Halliburton, Baker Hughes, Brown and Root, and Oceaneering became the major source of technological development through aggressive research and development programs. These firms continue to provide the infrastructure, equipment, expertise, and vendor services necessary for exploration and development companies to meet their production goals. Finally, offshore exploration gave birth to highly specialized logistical support and product transmission infrastructures required to transport hydrocarbons to market.

As petroleum exploration activity migrated into the wetlands, oil companies recruited local boat owners (usually Cajun fishermen) to transport small equipment, supplies, and workers to remote work sites. As the oil industry moved farther offshore, these logistical support companies were compelled to adapt to a new work environment. Over time, these Cajun mariners developed technologies critical to offshore logistics. Indeed, the coastal plain's workforce is responsible for numerous inventions and technological improvements used in the oil and gas industry, particularly offshore. Port Fourchon—Louisiana's southernmost seaport—is particularly noteworthy because of innovations that have made the compact site near the mouth of Bayou Lafourche the logistical nexus of the offshore oil industry in the northern Gulf of Mexico. This multiuse intermodal facility, established in 1960, features a broad spectrum of services ranging from dry docks to covered berths for a wide range of deepwater work vessels. The Fourchon facility, which has expanded over time to meet petroleum industry needs, now encompasses over twelve miles of bulkheaded waterfront dock space, and its slips can accommodate vessels over 300 feet in length. The port's client base, originally consisting of two companies, has grown to more than 250 businesses accounting for approximately 90 percent of the petroleum industry's infrastructure and transportation activities in the entire Gulf of México.

These companies were attracted to Fourchon by the port's vaunted reputation as a "one-stop shop." At Port Fourchon, offshore supply vessels can unload cargo, load supplies and equipment, and take on all necessary fluids from fuel to potable water with a turn-around time of about eight to ten hours—all usually without leaving their leased covered slip. The approximately 400 vessels that call on the port each day require prodigious quantities of supplies that are delivered daily by 1,200 semitrucks—a fleet roughly 10 percent the size of its counterpart at the massive Port of Long Beach.

Movement of bulk petroleum products from offshore wells also required highly specialized technological adaptations. Companies had to elevate tank batteries and service facilities by barge. Camp boats and living quarters had to be engineered and built. Offshore loading sites and pipelines had to be constructed and perfected by trial and error.

These pipelines, in turn, fed the sprawling petrochemical complex along the Mississippi River between Baton Rouge and Chalmette. This complex, frequently called "America's Ruhr," now boasts more than 100 existing or proposed plants. Most of these large petrochemical installations, which generally produce plastics and agricultural chemicals, opened after 1960, permitting the Port of South Louisiana to develop into one of the country's largest ports, handling more than 500 million tons annually. The concentration of these plants and refineries along this stretch of the Mississippi resulted from Governor John McKeithen's business-friendly tax policies (1960s); Governor Edwin Edwards's exceptionally lax environmental oversight (1970s–1990s); the abundance of inexpensive, locally available natural gas; a virtually unlimited supply of freshwater; a large pool of non-unionized blue-collar labor; and the riverside facilities' ability to export product directly to international and domestic markets by inexpensive waterborne transportation.

The continuing development of coastal Louisiana's nationally vital energy infrastructure in America's most vulnerable coastline has generated huge anticipated economic benefits at the expense of myriad, often unanticipated environmental, cultural, and social costs. How powerful private and corporate interests, governmental entities, and the local population collectively balance these competing costs will effectively determine the population's continuing ability to occupy the coastal plain and exploit its resources in an era of rapid climate and environmental change.

Beginning in the 1930s, operating engineers, using large bucket dredges, opened the wetlands to oil and gas exploration. (Photo courtesy of the Louisiana Department of Wildlife and Fisheries)

United Fruit Company

United Fruit Company docks, New Orleans. Samuel Zemurray founded the United Fruit Company, which evolved into the largest importer of Central American bananas. In the early twentieth century, the American banana trade was consequently centered in New Orleans. (Photo courtesy of the National Archives and Records Administration, Prints & Photographs Division, College Park, MD, photographer and date unknown, call number 4a19873a)

As a result of its massive overseas plantation and banana import operations, United Fruit Company wielded enormous political and economic influence in Central America, and New Orleans, as the corporation's home base, consequently functioned as the nation's principal window onto the Caribbean Rim until the Cuban Revolution of 1959. (From the authors' collections)

The United Fruit Company, based in New Orleans, distributed bananas—the world's fourth major food, after rice, wheat, and milk—from its plantation holdings in Central America to ports around the perimeter of the United States and elsewhere. As a result of this economic interaction, New Orleans was America's window to Central America until the Cuban revolution. (From the authors' collections)

Ice

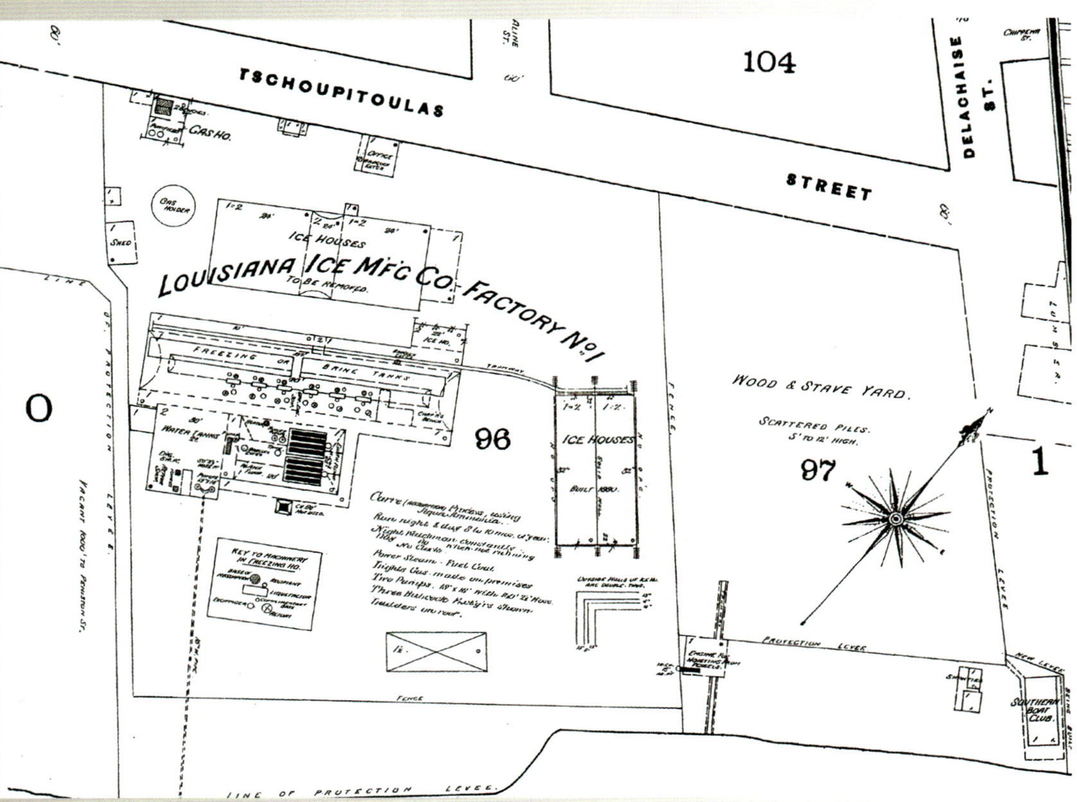

The Louisiana Ice Manufacturing Company was an important component of the Crescent City's ice industry. This plant, located behind the protection levee on the Mississippi River's east bank, probably served the fishing fleet servicing New Orleans markets. (Map from *Sanborn Fire Insurance Maps*, New Orleans, vol. 3, sheet 53 [New York: Sanborn Map and Publishing Company, 1887])

Oyster meats may have been canned first in New York City in 1819. By 1840, however, Baltimore had eclipsed its northern rival as the nation's premier oyster processor. In the succeeding decades, Baltimore oyster businesses shipped bivalves throughout the Midwest in cans packed in wooden cases lined with ice. This technology was also used in Louisiana, but on a more limited basis. (From the authors' collections, used with permission of Chris Cenac)

Donaldsonville

ICE FACTORY,

Mississippi Street,

H. COOK, - - - Proprietor.

WITH new machinery of most improved pattern, and enlarged facilities generally, am better prepared than ever before to supply the people of Ascension and adjacent parishes with ice, by the pound or block, at

Lowest Market Prices.

Having accepted the agency of the celebrated SOUTHERN BREWING CO., of New Orleans, and built a spacious REFRIGERATOR, I can furnish dealers and consumers with

Ice-Cold Beer by the Keg,

cheaper and in far better condition than it can be procured from New Orleans or elsewhere.

Have also purchased the steam propeller Harris Irvine and will make trips as follows, delivering ICE AND BEER to dealers and consumers along the respective routes, viz:

To College Point, Tuesdays and Fridays.
To Napoleonville, Wednesdays and Saturdays.
To Bayou Goula, Thursdays and Sundays.

Orders left aboard the boat, at the Factory or addressed to me through P. O. Box 32, Donaldsonville, will receive prompt attention.

Respectfully, H. COOK.

An 1886 advertisement for the Donaldsonville Ice Factory. (*Donaldsonville Chief*, July 10, 1886)

ICE! ICE!!

Mr. CHARLES WIECK having received a large stock of

PURE LAKE ICE

is now prepared to furnish the same to coast trade, and supply all local demands at reasonable rates.

All orders from Plaquemine, Bayou Goula, Port Hudson, Bayou Sara, Woodville, Jackson and Clinton promptly attended to.

CITY ICE HOUSE.

Open from 5:30 A. M., to 8 P. M. feb22

An 1879 Baton Rouge advertisement for imported lake ice. (*Baton Rouge Louisiana Capitolian*, March 22, 1879)

ICE DEALERS.

CRESCENT CITY ICE COMPANY, Office 21 Camp. J. F. Kranz, Pres't; J. J. E. Massicot, Secr'y and Treas.

SHRIEVEPORT ICE DEPOT, Cross Bayou, foot Market St., Shrieveport, La. C. H. Bosworth.

ICE MANUFACTURERS.

LOUISIANA ICE MANUFACTURING CO., Office 69 Canal; Works Tchoupitoulas St. near Louisiana Ave. Wm. T. Hepp, Pres't; D. Pochelu, Sec'y and Treas.

Ice factories were vital to the development of Louisiana's seafood industry. In the early twentieth century, locally produced ice permitted preservation of perishables in urban homes. (From the authors' collections)

Lumber

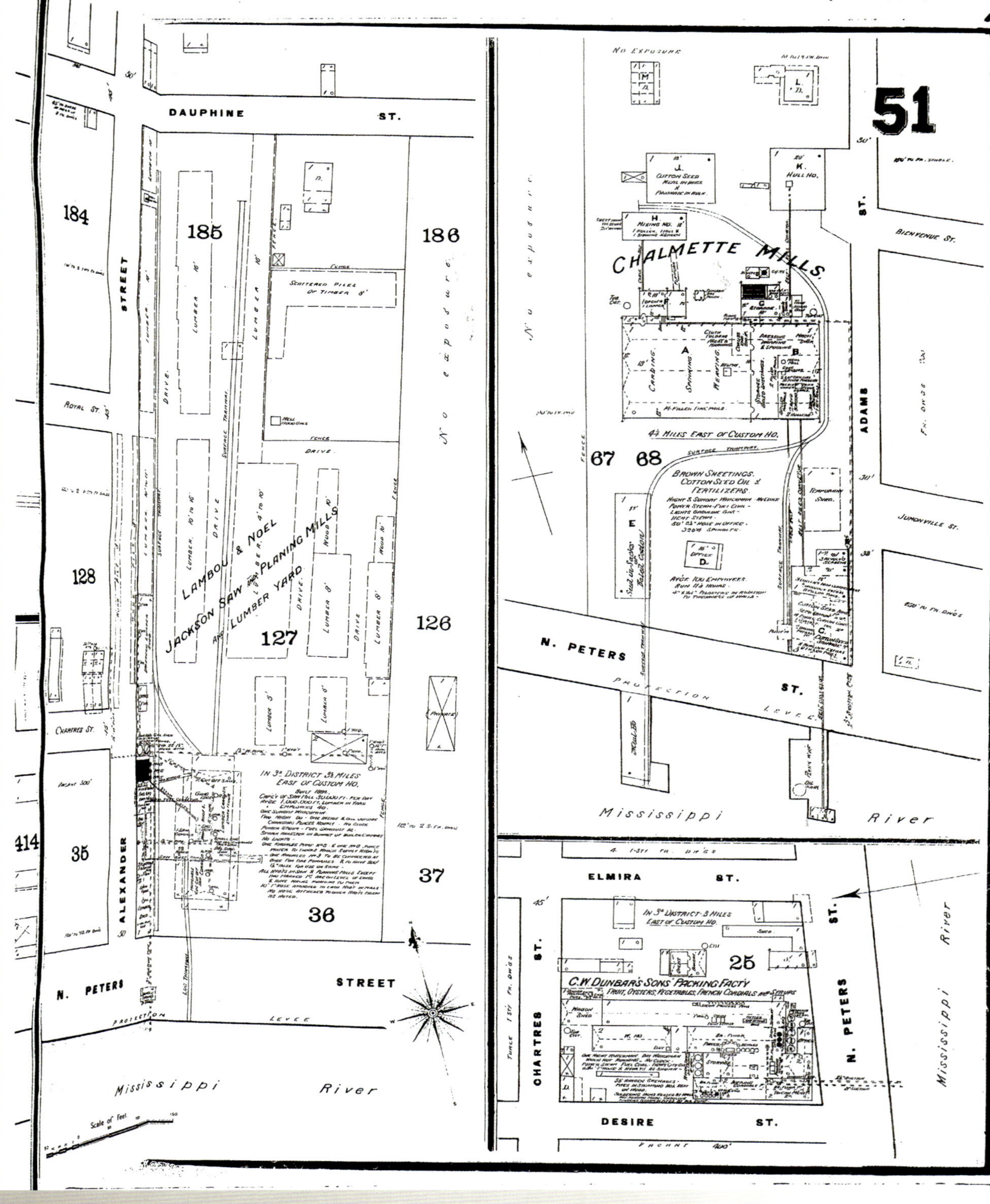

These maps provide detailed views of early industrial sites in New Orleans. The Lambou and Noel Jackson Saw and Planing Mills and Lumber Yard had a capacity of 30,000 board feet per day and a storage capacity of 1,000,000 board feet. The Chalmette Mills site boasts a large building for carding, spinning, and weaving cotton. The G. W. Dunbar's Sons Packing Facility canned fruit, oysters, vegetables, French cordials, and a variety of syrups with steam-powered equipment. (Map from *Sanborn Fire Insurance Maps*, New Orleans, vol. 31, sheet 48 [New York: Sanborn Map and Publishing Company, 1887].)

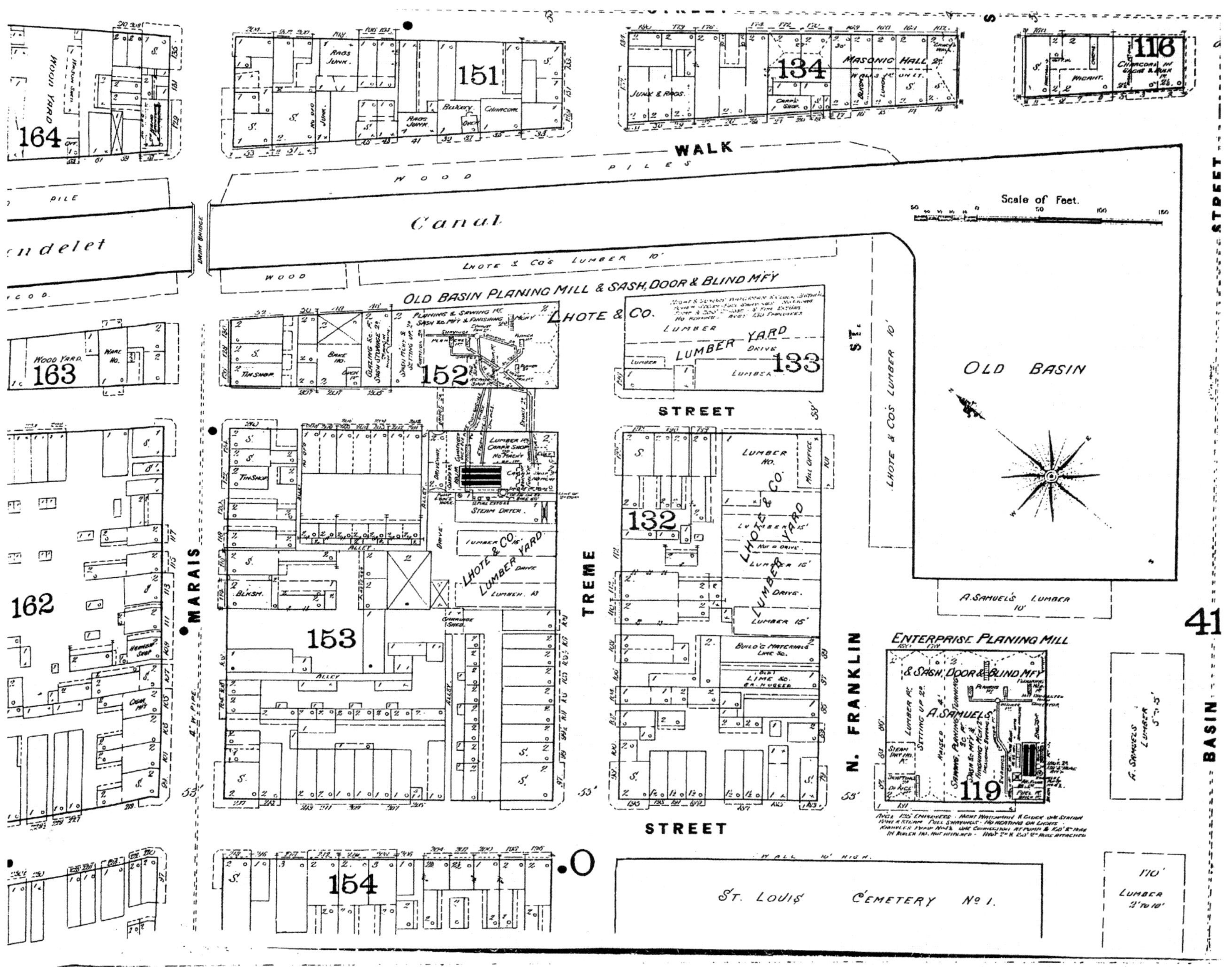

The city of New Orleans filled in the New Basin Canal during the 1950s to create the Pontchartrain Expressway and West End Boulevard. The canal originally emptied into Lake Pontchartrain near the Southern Yacht Club. Completed in 1838 by Irish immigrants, more than 8,000 of whom died during the project, the canal connected the downtown American sector with the lake. By the late 1800s the channel was lined with a variety of lumberyards; planing mills; sash, door, and blind companies; firewood yards; cooperages; and a vast assortment of other ancillary businesses. (Map from *Sanborn Fire Insurance Maps*, New Orleans, vol. 2, sheet 25 [New York: Sanborn Map and Publishing Company, 1887])

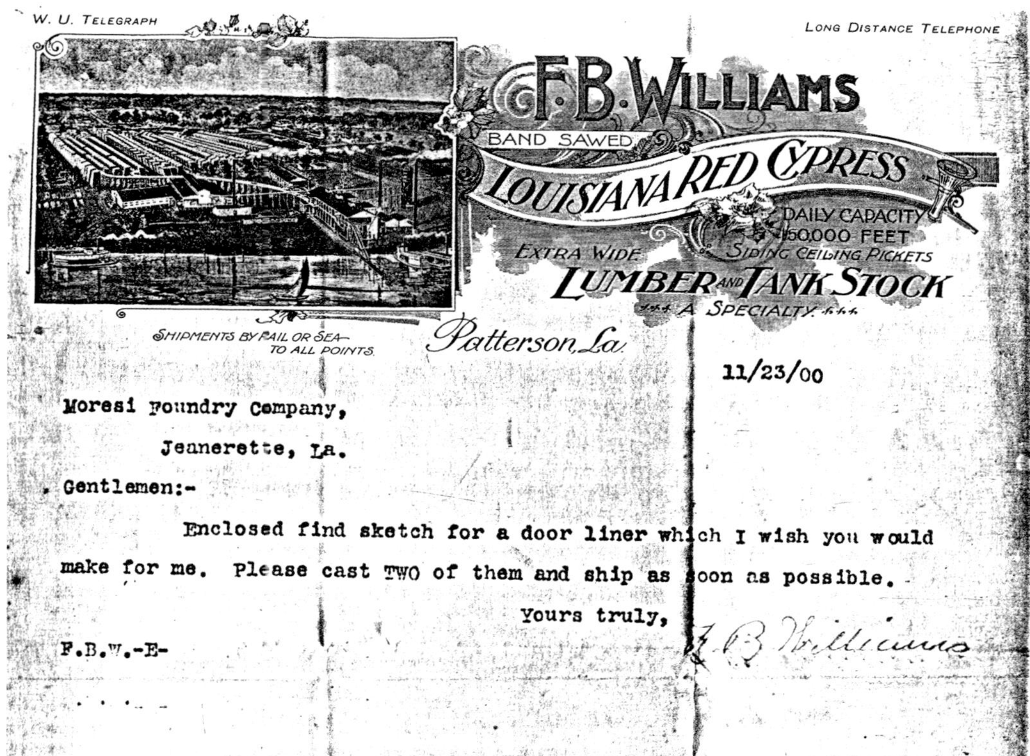
W. U. TELEGRAPH

LONG DISTANCE TELEPHONE

F. B. WILLIAMS

BAND SAWED

LOUISIANA RED CYPRESS

DAILY CAPACITY 150,000 FEET

EXTRA WIDE

SIDING CEILING PICKETS

LUMBER AND TANK STOCK

A SPECIALTY

SHIPMENTS BY RAIL OR SEA— TO ALL POINTS

Patterson, La.

11/23/00

Moresi Foundry Company,

Jeanerette, La.

Gentlemen:-

Enclosed find sketch for a door liner which I wish you would make for me. Please cast TWO of them and ship as soon as possible.

Yours truly,

F. B. Williams

F.B.W.-E-

As the stationery's masthead makes abundantly clear, cheap and reliable transportation was critical to the mills' success; hence, milling operations were invariably located along streams or spur railroad lines. (Courtesy of the Morgan City Archives, Morgan City, LA, 1900)

Cable Address "MAY"

Building Toward A Bigger - Better Louisiana

MAY BROTHERS

GARDEN CITY, LOUISIANA, U.S.A.

MANUFACTURERS

HARDWOOD LUMBER AND LOUISIANA RED CYPRESS

DISTRIBUTORS OF ESSENTIAL PRODUCTS TO THE OIL INDUSTRY

CHEMICALS - MAYGEL - MAYBAR - DRILLING MUDS - CEMENT

PHONES: GARDEN CITY 1750 — MORGAN CITY 3340 AND 6491

The May Brothers mill at Garden City was the economic centerpiece of a community that remains one of Louisiana's best-preserved former company towns. (Photo courtesy of the Morgan City Archives, Morgan City, LA, date unknown)

The steamboat *Norman*, towing a raft of cypress logs to a Plaquemine sawmill, 1928. (Photo courtesy of the National Archives and Records Administration, Prints & Photographs Division, College Park, MD, photographer unknown, call number RG 54-Y, Box 1, Y-938)

Shrimp and Oysters

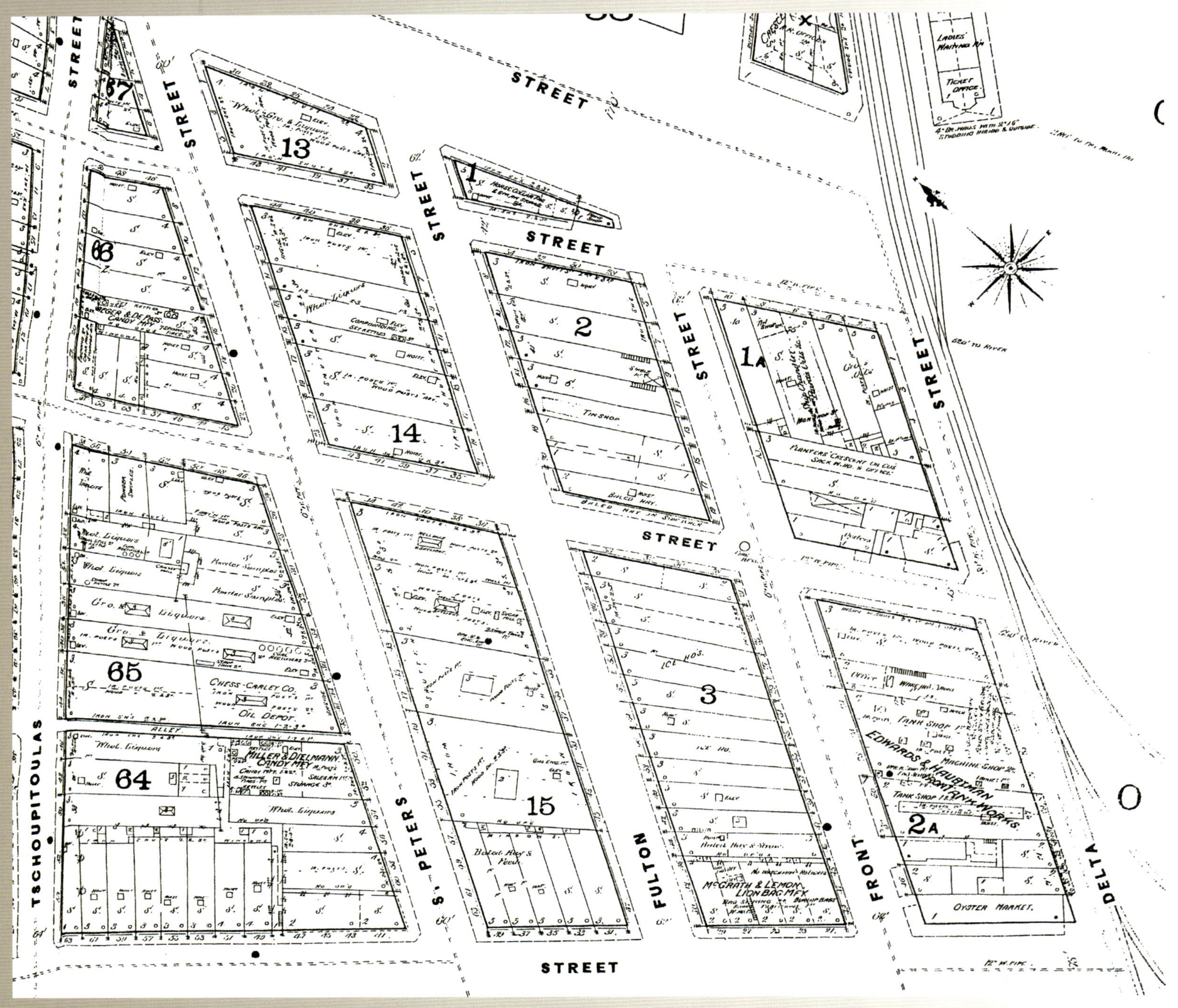

Note the number of icehouses supporting the oyster market facing Lafayette Street, between Front and Delta Streets. The size of the ice storage area is an indication of the importance of oysters to the city's economy. Icehouses used sawdust, obtained from area sawmills, as an insulator. (Map from *Sanborn Fire Insurance Maps*, New Orleans, vol. 2, sheet 19 [New York: Sanborn Map and Publishing Company, 1887])

The earliest shrimp-drying platforms were supplied by seine crews that lived at the sites. In this case, the vessel *Bulldog*, with at least thirty barrels of shrimp, supported the seine crew. (Photo from Tate Album, Mss. 4963, LLMVC, photo number 4963058)

Prior to refrigeration, scores of isolated, short-lived shrimp-drying platforms sprang into existence throughout southeastern Louisiana's coastal marshes. (Courtesy of the Morgan City Archives, Morgan City, LA, photographer Jesse Grice, date unknown)

A CREOLE COOK-BOOK

"COOKING IN OLD CREOLE DAYS"

By CÉLESTINE EUSTIS

This book gives recipes for all of the famous old Creole dishes—many of them having never before appeared in print—and explicit but clear directions are given for their preparation. The book includes as well a number of quaint old Creole songs in praise of famous dishes. The recipes are also given in French.

Charmingly Illustrated. Decorative Paper Sides, Cloth Back, $1.50

(*Imprint of R. H. RUSSELL*)

HARPER & BROTHERS, PUBLISHERS, FRANKLIN SQUARE, N. Y.

Creole cookbooks emanating from the Crescent City helped create a national demand for Louisiana seafood. (Lithograph from *Harper's Weekly*, December 9, 1905, 1792)

Early twentieth-century Louisiana shrimpers showing the equipment used for moving their catch to processors. This equipment included "chinee" baskets. Derived from containers evidently introduced into the state by Chinese platform shrimpers, these baskets constituted the standard measure for commercially marketed shrimp. (Photo courtesy of the National Archives and Records Administration, Prints & Photographs Division, College Park, MD, photographer and date unknown, call number RG 22-FC, Box 1, subset-FCD, photo number CD 227)

Directly from the Fishermen to You

We have the Shrimp you need to satisfy your growing, thriving markets. We're an organization of independent boat owners and fishermen, and we offer you a fresh, prompt supply.

We solicit business from old line dealers, and are not interested in consignment shipments. Yes, if you want fast delivery of firm, quality shrimp, and you're willing to cooperate by paying promptly - - - we want to do business with you. Get in touch with us by mail, wire or phone.

As the Fleet is Blessed, we add our voice to the multitude of those who wish nothing but good to the fishermen during the coming season.

★

FISHERMEN'S COOPERATIVE ASSOCIATION, INC.

SHRIMP

MORGAN CITY, LA.

In the early twentieth century, fishermen and trappers in the Morgan City area felt compelled to organize to combat alleged exploitation by their respective industries' middlemen. (Advertisement courtesy of the Morgan City Archives, Morgan City, LA, date unknown)

Although their nets are often made from monofilament webbing, Louisiana fishermen continue to treat their nets with a tarlike substance. This is a tradition that goes back to a time when cotton was the fiber of choice for fishing nets. (Photo by authors, 2015)

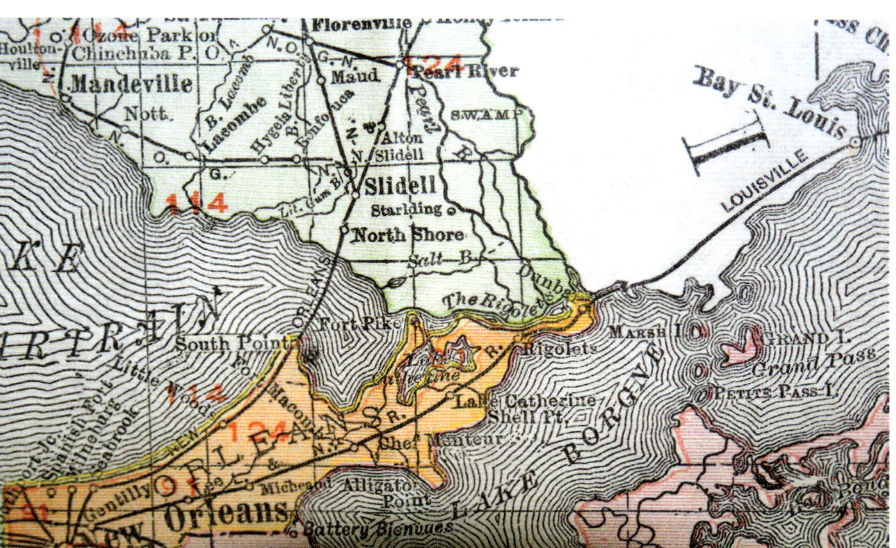

Around 1890 the Dunbar canning company occupied and renamed the community of Lookout (sometimes dubbed English Lookout). (Map from Rand McNally's *Indexed Atlas of the World, Louisiana* [Chicago: Rand, McNally, 1892])

Stockyards

These custom-built barges were designed specifically to transport cattle through the most treacherous portions of the coastal wetlands cattle trails. (From the authors' collections)

Southwestern Louisiana's cattlemen have traditionally preferred locally adapted hybrids to purebred strains because of their ability to thrive in harsh wetlands conditions and their superior resistance to indigenous pests. (Photo by the authors, 2010)

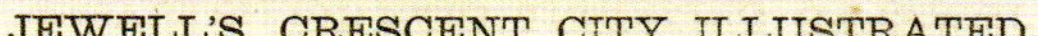

JEWELL'S CRESCENT CITY ILLUSTRATED.

THE SLAUGHTER HOUSES.

ABOUT four miles below Canal street, immediately above the dividing line of the parishes of Orleans and St. Bernard, and about two hundred yards below the United States Barracks, stands the aggregation of buildings of which a partial view is herewith exhibited. This is the spot where the law prescribes that all animals intended for our markets must be slaughtered, under the supervision of officials appointed, some by the State and others by the "Crescent City Live Stock Landing and Slaughter House Company,"

The arrangements for carrying out the objects contemplated by the law creating the Slaughter Houses are very complete. There are two wharves for landing the stock, with pens for receiving them upon the Levee. Immediately adjoining the wharves, and also upon the Levee, is a building containing a pumping apparatus worked by steam and capable of supplying 150,000 gallons of water daily to the entire establishment. Fronting the public road, a row of one-story buildings contains the telegraph, the company's offices and also the private offices of Messrs. C. Mehle & Co., Inbau, Aycock & Co., and L. B. Collins, live stock dealers.

In the rear of these offices at 200 feet from the public road, there are twelve large covered cattle pens, 67 by 15 feet each, where the live stock are first placed for inspection and sold to the various stock dealers who are always in attendance. Adjoining these are the main pens for cattle, twenty-eight in number, and each one having the dimensions of 75 feet by 17, and also eighteen hog pens for sheep and other cattle. After these you come to the large hog and sheep slaughter house, 265 by 80 feet. which is constantly supplied with hot and cold water, pulleys, etc. On the right hand as you look toward the river, is another immense building containing twenty-two divisions 32 by 25 feet for slaughtering beeves and calves, each division having in its rear two large pens 60 by 10 feet for receiving the cattle previous to slaughtering. Fronting the above building from which they are separated by a broad paved alley for carts are twenty-two stables for the horses of the butchers, and next to these, nearer to the river, are covered vats for salting and curing the hides. Some fifty feet in the rear of the whole is another steam engine, and a little further back of this stands the Blood Fertilizer Manufactory, occupying a space of 200 feet square, lodging horses for the employèes etc. The outhouses are all new and in perfect order, and in another part of the grounds there is an apparatus for curing hides by acids in two hours, and Esteban's large sheep-skin tannery. The other buildings are rented by the company to various parties, who are principally coffee-house and tavern keepers. The charges for slaughtering are as follows:

Beeves	$1 00 per head.
Calves	50 "
Hogs	50 "
Sheep	30 "

The average number of cattle slaughtered daily is about 700 during the summer, and from 900 to 1,100 in the winter. With the present arrangements there is ample accommodation for the slaughter of 1,500 head of cattle daily, and these facilities can be increased to any extent, the

The New Orleans slaughterhouses jettisoned offal into the Mississippi River, and, during seasonal "low-water" phases, "intestines and portions of putrefied animal matter" clogged municipal drinking pipes, contributing directly to sporadic deadly cholera outbreaks. Following the Civil War, the city's government filed suit—the nationally famous Slaughter House Case—against the local meat processing industry to force it to mitigate this public health hazard. As the case wended its way to the US Supreme Court, the suit somehow morphed into a civil rights case based on the recently ratified Fourteenth Amendment. The controversial benchmark decision, which temporarily eviscerated the amendment, is now considered one of the most notorious rulings in American jurisprudence. (From the authors' collections)

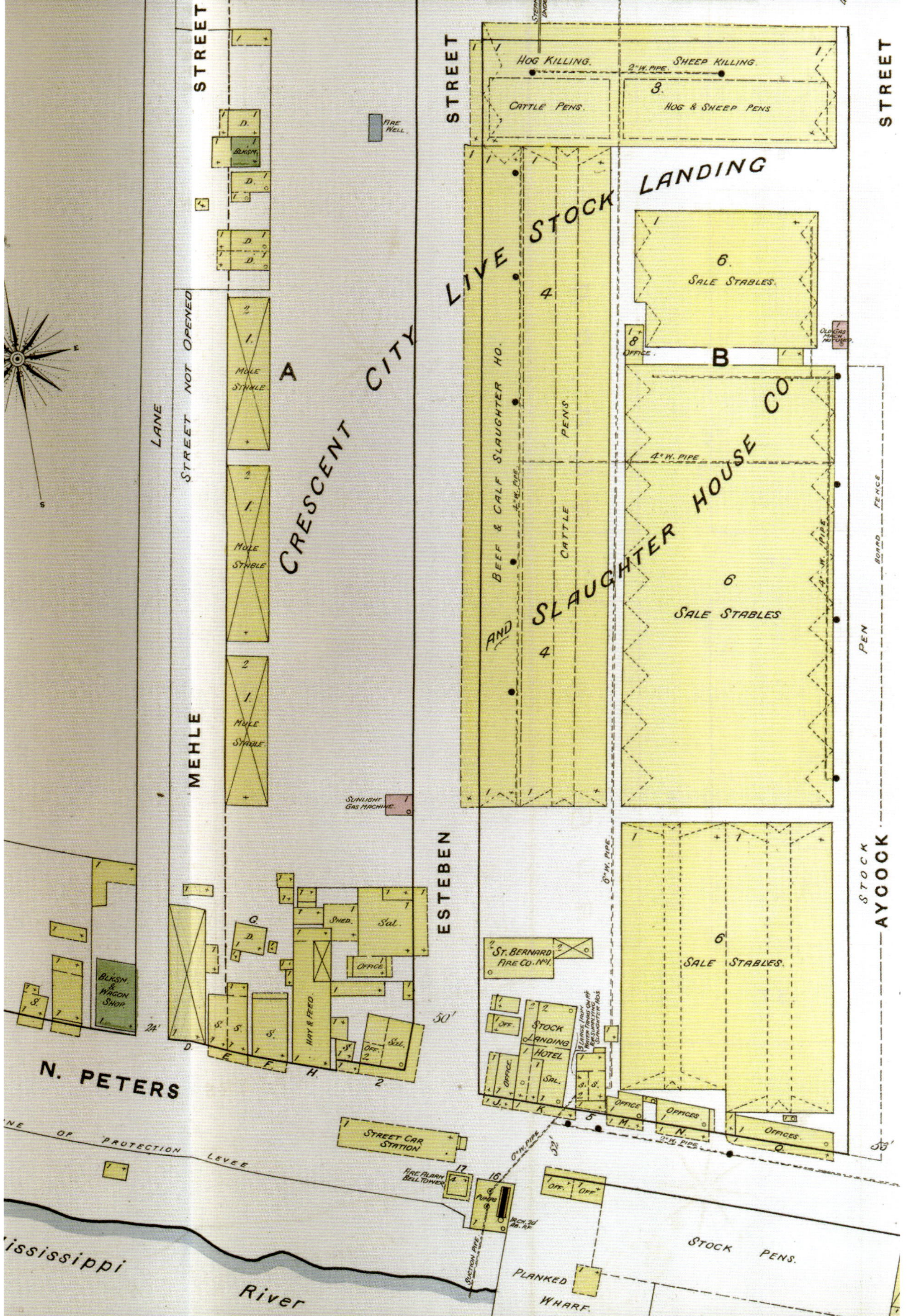

In the mid-nineteenth century, thousands of butchers annually processed 300,000 beef carcasses at slaughterhouses upstream from New Orleans. The discarded intestines, dung, blood, and urine contaminated the city's drinking water, causing at least eleven cholera outbreaks. To address this issue, the Louisiana legislature chartered a private corporation to run a "Grand Slaughterhouse" at the southern boundary of the city, now the city of Arabi. By statute, all animals destined to be slaughtered were to be landed at this facility's stock landing and nowhere else. (Map from *Sanborn Fire Insurance Maps*, New Orleans, vol. 3 [New York: Sanborn Map and Publishing Company, 1887])

On the south shore of Lake Pontchartrain, along Haynes Boulevard, the local levee board issued permits allowing construction of elevated camps connected to the shore by narrow, elevated walkways. These camps were to New Orleans what beaches are to Florida; however, after Hurricane Katrina, this community of lake dwellers largely disappeared. (Photo by the authors, 2012)

4

CONCLUSION

Coastal Louisiana has undergone a net change in land area of about 1,883 square miles (mi^2) from 1932 to 2010. This net change in land area amounts to a decrease of about 25 percent of the 1932 land area. Persistent losses account for 95 percent of this land area decrease; the remainder are areas that have converted to water but have not yet exhibited the persistence necessary to be classified as "loss." Trend analyses from 1985 to 2010 show a wetland loss rate of 16.57 mi^2 per year. If this loss were to occur at a constant rate, it would equate to Louisiana losing an area the size of one football field per hour.

Source: B. R. Couvillion, J. A. Barras, G. D. Steyer, W. Sleavin, M. Fischer, H. Beck, and D. Heckman, *Land Area Change in Coastal Louisiana (1932 to 2010)*, US Department of the Interior, US Geological Survey (2011), 1.

The industrialization of the Louisiana coastal plain, which the sugar mills, shrimp docks, oil canals, petrochemical complex, the region's fabrication operations and shipyards, and massive oil service facilities lining "America's energy corridor" (Highway 90 between Lafayette and Houma) graphically represent, has completely transformed not only the coastal plain's economy but also its human and natural landscapes. The metamorphosis afforded area residents the promise of a better life—or at least a higher standard of living—at the cost of far-reaching social, cultural, and environmental change that threatens not only the established population's cherished traditional way of life but also its continuing ability to occupy the wetlands that have defined their collective identify for more than two centuries.

The pace and magnitude of the coastal plain's social, economic, and environmental transformation have been stunning. A marshland resident magically teleported from the early 1980s to the present would scarcely recognize his or her native wetlands. Although change is inevitable—indeed, the only constant—the path of the region's evolutionary track has been profoundly shaped by the so-called stakeholders'—local, state, and national—marked collective propensity for trading short-term gains (often for the privileged few) at the cost of long-term, frequently intractable problems (for the many).

This mindset first crystallized in Louisiana's formative colonial origins. After 1717 the focus of French Louisiana's development shifted from Biloxi and Mobile to the lower Mississippi River valley. The first colonial settlements were established on the natural levees, which, while permitting exploitation of some of the world's most fertile soil, also exposed settlers to catastrophic seasonal flooding and hurricane damage. Settlers sought to counter the threat of natural disasters by attempting to mitigate the most immediate threat—riverine flooding—by building crude levees (earthen embankments) along the

Representative examples of short-term gains at the expense of long-term problems include:

- Privatization of the coastal wetlands, which for generations had been a shared public resource, put the region's future in the hands of corporate interests—often absentee landlords—preoccupied with remunerative short-term returns.
- The introduction of invasive species, the most egregious example of which is introduction of the nutria (*coypu*) to further the short-term economic interests of landowners seeking increased revenues from marshlands leased to trappers.
- Using the wetlands as waste disposal sites without considering the long-term consequences of these often unregulated practices.

Proclamation by Alejandro O'Reilly, Spanish governor of Louisiana, February 18, 1770: "The concessionaires established along the river will be required, during the first three years of occupancy, to build levees sufficient for the protection of the properties and ditches necessary for drainage."

Source: Archivo General de Indias, Seville, Spain, Papeles Procedentes de Cuba, legajo 189B, folios 1–4.

riverbanks. The French colonial government (1699–1763) and its Spanish successor (1763–1803) encouraged these bulwarks. In fact, Spanish land grant regulations promulgated on February 18, 1770, mandated the construction of a modest river- or bayou-side levee as a major condition for issuance of royal patents. Governmental mandates for private levee construction and maintenance became increasingly stringent after the American acquisition of Louisiana in 1803, as plantation owners used their considerable political influence to protect their economic interests by means of engineered earthworks.

Although feeble in comparison to the massive earthen structures presently lining the Mississippi and its major distributaries, this primitive levee system nevertheless provided a modicum of protection to the farms and towns along the region's navigable waterways. This protection, however, came at the cost of a fundamental change in the natural hydrology that built and sustained this ecologically fragile biome.

These environmentally detrimental practices not only persisted but actually expanded exponentially in the late nineteenth century, as well-heeled agriculturalists secured establishment of the Mississippi River Commission in 1879. Charged with improving "the conditions" of the Mississippi River, fostering navigation, promoting commerce, and preventing destructive floods between Louisiana's Head of Passes and Cape Girardeau, Missouri, the commission mobilized for the first time national resources to alter the great river basin's natural hydrology.

The prolonged effort to control the river's natural flow regimes reached its climax in the late nineteenth and early twentieth centuries. Responding to complaints from planters about the perennial flood threat over the course of at least five decades, police juries (the Louisiana counterpart to county commissions) and levee district boards authorized and underwrote the damming of the two Mississippi distributaries most responsible for widespread flooding in the coastal plain's natural levee farming communities—Bayous Plaquemine (1866) and Lafourche (1904).

Construction of an earthen dam across the head of Bayou Lafourche at Donaldsonville proved particularly problematic, because it interrupted the remaining flow of waterborne nutrients and sediments from the Mississippi into the Barataria-Lafourche-Terrebonne estuarine complex—one of the world's most biologically rich and productive ecosystems. The resultant regional environmental problems, however, pale by comparison to those caused by the massive "replumbing" of the lower Mississippi Valley's natural hydrology through construction of the present levee system following the devastating 1927 flood.

Through the Flood Control Act of 1928, the enabling legislation for this massive engineering project, the federal government committed itself to a flood control program that would create and maintain the world's longest levee system. Implementation of this program was immediate and rapid, and by 1936 the Mississippi had twenty-nine locks and dams, hundreds of runoff channels, and at least 1,000 miles of levees. The river, for the first time, was effectively confined by an engineered straitjacket.

These public flood-control efforts were complemented by numerous large-scale private land reclamation campaigns designed to transform pristine wetlands into productive farmlands or, in the New Orleans area, suburban developments. For example, the Northern States Citrus and Realty Company, the driving force behind creation of a southern "Citrus Belt," played a pivotal role in transforming the marshy Lake Pontchartrain shoreline into the community of Citrus in present-day New Orleans East.

In rural areas, reclamation developers, who aggressively promoted their projects through the Louisiana Reclamation Club, extolled the great economic opportunities that wetland reclamation afforded immigrants willing to transform "unproductive wastelands" into thriving agricultural enterprises. These rural reclamation developments coincided with the political triumph of agricultural and industrial interests over rival enterprises—particularly ranching.

Since the late eighteenth century, cattlemen had utilized southwestern Louisiana's great prairie lands and natural ridges in the coastal marshes as open ranges with great success. As agricultural settlements expanded beyond the confines of their original colonial boundaries during the nineteenth century, farmers—particularly sugar planters—increasingly came into conflict with cattlemen, whose at-large animals not only posed a threat to plantation crops but whose continued use of public domain lands jeopardized their ability to expand their property holdings—and profits. Around the turn of the twentieth century, agricultural lobbyists prevailed upon the Louisiana legislature to adopt legislation imposing fencing requirements upon ranchers, thereby effectively confining the traditional cattle industry to the upper margins of the coastal marshes.

Agriculture's political triumph roughly coincided with the rapid settlement and economic development of the prairie region. In 1880–1881 the Louisiana Western Railroad Company's (later the Southern Pacific Corporation) gandy dancers laid the last section of track for the trunk line extending from Vermilionville (now Lafayette)

The Atchafalaya River, which had also been a major irritant for planters on both sides of the basin since clearance of a ten-mile-long logjam in 1842 unleashed its flooding potential, proved a much more formidable challenge. The Atchafalaya River raft reformed in 1846, and the Louisiana government initiated a new logjam removal project in 1858. Real control of that drainage system, however, would not be achieved until completion of the Morganza Spillway system in 1954 and the Old River Control Structure in 1963.

"After the great 1927 flood, Congress proposed a project to build along the lower Mississippi from Cairo, Illinois[,] basically to the Gulf of Mexico to ensure that they never have a[nother] 1927 flood in the Mississippi. They assigned the overview and administration of that project also to the Mississippi River Commission. . . . It has been going on since 1928. It's probably appropriated, or spent, about $13.5 billion. It has prevented many floods along the river. It has been designed to prevent the maximum flood, which [the] 2011 [flood] has probably been the closest thing to that in the last eighty years."

Source: Clifford Smith, interview by authors.

Recruitment bureaus funded by reclamation and railroad companies portrayed the Louisiana coastal plain as an earthly paradise, featuring mild summers and winters, bountiful soil, and friendly, if backward natives. One recruitment brochure, for example, claimed that lower Louisiana "lacks the killing heat of the northern latitudes. Sunstroke is an unknown malady. The winters are never severe, as the mercury seldom sinks to the freezing point. . . . About once in seven or eight years there is freezing weather."

Source: West Louisiana, East Texas, and Gulf Coast: Along the Kansas City Southern Railroad (Beinecke Library, Yale University).

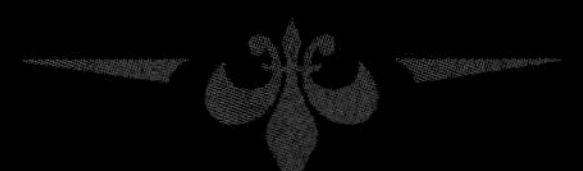

These [upper coastal] marshes that we're driving through here now, were all productive cattle areas and still are. There's cattle grazing all over this prairie, up here, and this is kinda the interface between high land that extended south of DeQuincy down to Lake Charles/Sulphur area, and then tapered off into the marsh itself.

Source: Allen Ensminger, interview by the authors.

It was properties that they had drained [that experienced] subsidence and oxidation. Once you dry these marshes up they just turn into a sponge bed. And it happened there with the Netherland Corporation. They drained it. [The] 1915 hurricane came through there and flooded it, went over top of those protection levees.

Source: Allen Ensminger, interview by the authors.

to Houston. The railway's completion sparked an explosion of land-speculation activity along the railroad's right-of-way. Some speculators, such as brothers C. C. and W. W. Duson, focused on town development, while others, primarily out-of-state transplants, capitalized a massive recruitment campaign that resulted in a large-scale influx of midwestern farmers anxious to flee the rigors of northern winters. Most of the migrants settled on the formerly open prairie lands, but a significant minority of the immigrants initially set down roots on lands opened to the public as short-lived reclamation projects that were often lauded as "New Netherlands."

After a failed experiment with wheat production, these transplants turned to rice cultivation. Large-scale rice output, of course, required the establishment and maintenance of one of the world's largest commercial irrigation systems, and, while some of the essential irrigation water was drawn from shallow wells, most was generated through a well-designed network of low, narrow diversion projects, often with significant environmental consequences. For example, during the decades bracketing the turn of the twentieth century, wealthy rice interests reaped the benefits of the state government's benign neglect by damming the lower Mermentau River—at a cost of $75,000—to provide additional *public* water to *private* commercial irrigation systems. The earthen barrier launched a decadelong de facto war, as enraged residents downstream successfully used dynamite repeatedly to restore the natural hydrology after equally dogged commercial efforts to maintain the hydrologic modifications.

The short-term effects of such environmental upheavals were not immediately apparent, but the long-term consequences have proven inescapably obvious, even for the most committed skeptics. Sediment-deprived lands naturally subside over time and, without vital waterborne nutrients to renew the native plants sustaining the "trembling prairies" that constitute the living fabric of the coastal wetlands, consequently became stressed. Over time, stressed "coastal carpets" become increasingly susceptible to storm surges, cold-front passages, and daily tides, particularly after commercial and industrial canals destroyed the integrity of these luxuriant floating carpets. The magnitude of the dredging operations alone is hard to overstate. For example, between 1900 and 2000 industrial dredges excavated access canals to service more than 67,000 well sites, in the process destroying the critical structural integrity of the coastal marsh's floating prairie.

During construction of these canals, dredges piled organic material and sediment into linear mounds (effectively levees known locally as spoil banks) that further disrupted the natural hydrology required to maintain the health of the swamps and coastal marshes. Nutrient-rich waters filtering through these wetlands not only

The news report below provides a snapshot of the dispute: "The trouble over the dam in the Mermentau at Grand Chenier, by the means of which the rice farmers along the river hoped to keep the saltwater from the gulf from ruining their crops is of several years standing. The construction of the dam was closely followed by a disastrous spring flood which overflowed much land adjacent to the dam, and the . . . land owners claimed that the flood was the result of the building of the dam.

"Shortly afterward a mysterious explosion resulted in the partial destruction of the dam. The rice men at once raised more money, had the dam repaired and placed guards at the structure. Last October the dam was again blown up by unknown parties. As the season has been wet, the dam was not needed this summer to protect the rice lands, and it is hoped to reach some adjustment of the dispute before another crop season opens."

Source: Crowley Signal, September 19, 1908.

In a recent interview with the authors, Houston Foret pointed to areas of open water at Cocodrie (lower Terrebonne Parish) and recalled a very different landscape only a few decades earlier: "Right here, there was a sugarcane plantation. Out there was a sugarcane plantation. They were raising corn over there. You know, the other [east] side of the bayou was the same, but there was [also] a little [sugar] refinery which was before my day. . . . And everybody had cattle. All the people raising or farming or whatever they would do."

This image—produced on September 15, 2004—is a computer-generated composite of Hurricane Ivan, dubbed "Ivan the Terrible" by the media. This storm shut down 25 percent of the Gulf's offshore oil industry for a month. The total damage from this storm was estimated at $18.8 billion. Hurricanes have traditionally constituted the greatest challenge to coastal residents; other examples include: Charley, Frances, Jeanne, Dennis, Katrina, Rita, Gustav, Ike, Irene, and Sandy. In recent times, these challenges have become increasingly formidable as coastal erosion continuously increases the destructive potential of storms making landfall between the Pearl and Sabine Rivers. (Image courtesy of NASA from http://rsd.gsfc.nasa.gov/goes/pub/goes/c40915.ivan.jpg. 2004)

sustained the area's indigenous flora and fauna, but they also provided sediments necessary to maintain the fragile landscape. With the modification of the natural hydrology, landforms began to subside, ecosystems began to collapse, and coastal wetlands—now shredded by thousands of canals—began to erode at an alarming rate as a result of enhanced saltwater intrusion and increased tidal action upon the floating prairies.

Within a matter of years, land-water ratios changed dramatically: canals grew into lakes, and lakes expanded into bays. The Barataria-Lafourche-Terrebonne estuary is now the national poster child for coastal erosion, accounting for about half of the state's wetland loss. Since the early 1930s, the Barataria-Terrebonne basins have lost at least 20 percent of their wetlands at a rate of eighteen to twenty-two square miles a year, resulting in the disappearance of a landmass about the size of Los Angeles and Indianapolis combined. Between 1956 and 1978 alone, more than 294,000 acres of marsh (460 square miles—four times the area of Staten Island) became open water, and coastal studies researchers suggest that one-third of the remaining wetlands could be lost by 2040.

But this unfolding environmental cataclysm is not confined to the lower Lafourche Basin. Since 1930 Louisiana's coastal parishes have collectively lost a landmass equal in size to the state of Delaware. No coastal parish has been spared, and the problem promises to escalate exponentially as subsidence and rising ocean levels place ever greater swaths of coastal wetlands at a heightened risk of washing away—either gradually through modest incremental losses or rapidly through catastrophic hurricane storm surges.

Disintegration of South Louisiana's once seemingly impenetrable and indestructible marshes has proven catastrophic for littoral parishes, whose topography renders them particularly susceptible to "high water" episodes. Clifford W. Smith Sr., dean of coastal Louisiana's civil engineers with nearly six decades of experience and a former civilian member of America's Mississippi River Commission, notes that his native Terrebonne Parish encompasses approximately 1.3 million acres. However, only about 5,000 acres (.0038 percent) are at least five feet above sea level. Such almost universally low-lying areas are increasingly vulnerable to the destructive action of hurricane-induced storm surges, such as those associated with Katrina (2005), Rita (2005), Gustav (2008), Ike (2008), and Isaac (2012).

The great stretches of open water that have replaced the living carpets of native grasses significantly facilitate ever deeper incursions of the Gulf's saltwater into the interior. Gum Cove Ranch, situated atop a modest ridge straddling the

The Mississippi River flood of May 2011 was among the largest and most damaging in recent memory. The Atchafalaya Basin flooded up to twenty feet, and more than 4,600 square miles were impacted by floodwater. Twelve of the Morganza Spillway's 125 floodgates were opened to reduce the Mississippi's flow past Baton Rouge. The opening of the floodgates on May 14, 2011, constituted the second time the structure was used to divert floodwaters. (Photo by the authors, 2011)

Cameron-Calcasieu line approximately sixteen miles from the Gulf shoreline, is a representative example. The original Gum Cove ranch house, constructed in the 1830s, survived scores of major hurricanes—including the catastrophic 1867 and 1957 storms—without "getting its feet wet," thanks to the ability of the adjacent marshes to tame even mountainous storm surges. As a result of wetland loss, accelerated by marshland scouring caused by Hurricane Rita in 2005, the area's current benchmark storm, the Gum Cove ranch house sustained severe flooding during Hurricane Ike in 2008. Even Lake Charles, a city of approximately 72,000 residents thirty-two miles from the Gulf, experienced notable flood damage in 2005 and 2008. Intensive post-Katrina scientific research, however, suggests that these flooding events are but precursors to far greater catastrophes. As Louisiana's natural coastal defenses crumble, the frequency and severity of storm-driven inundations will inevitably increase significantly, particularly in the face of rising sea levels.

This environmental degradation has ultimately threatened unique coastal communities and an attendant way of life more than two centuries in the making, the nation's most productive fishery, and a vital natural hurricane buffer protecting a $1 trillion, rapidly expanding petrochemical infrastructure that is vital to America's energy independence and economic security. For the first time since the 1893 hurricane, which claimed at least 2,000 lives, Louisiana's coastal population is in retreat. Plaquemines, St. Bernard, and Cameron Parishes have collectively sustained notable population

After Hurricane Katrina ravaged New Orleans and the surrounding communities, some residents felt compelled to tell the world that they would "rebuild better than ever." Despite such optimism, areas outside the Crescent City have been slow to rebound. In 2015—ten years after the storm killed more than 1,830 people—efforts to protect outlying communities have been slow to develop, and the loss of precious marsh buffers continues apace. New Orleans itself has become increasingly more vulnerable. (Photo by the authors, 2005)

In the 1930s the local school system floated two school buildings down Bayou Terrebonne to create the Indian School in Montegut. Although they do not enjoy official Bureau of Indian Affairs recognition, the tribal groups at Isle de Jean Charles and Point-aux-Chenes, south of Houma, are proud of their heritage and long historic attachment to the wetlands. They have been asked to move from their "homelands," which amounts, in their view, to cultural suicide. (Photo by the authors, 2000)

Debris left behind at Holly Beach by Hurricane Rita, September 2005. Because of stringent new construction requirements, most former campsite owners have sold their properties—often to Texans, who view the beachfront community as a much cheaper alternative to Galveston. (Photo by the authors, 2005)

losses. In fact, so many St. Bernard Parish residents have opted to relocate in St. Tammany Parish, on Lake Pontchartrain's northern shore, that the area has recently come to be known—only half in jest—as St. Tammanard.

In Cameron Parish, on the other hand, much of the remaining population has retreated to highways traversing, east to west, the region's northernmost frontier. In Terrebonne Parish, coastal residents are congregating behind the Morganza to the Gulf levee project, which, when completed, will extend ninety-eight miles from Gibson to Lockport. The western extension at Gibson is about thirty-five miles inland, while the eastern anchor point at Lockport is about twenty miles from Terrebonne Bay. Communities beyond the protection levee are quickly morphing into virtual ghost towns in which homes are replaced by recreational camps. Cocodrie, once a thriving community of 400 permanent residents, is now home to less than ten year-round inhabitants.

Retreat has also begun in ostensibly less vulnerable areas of the coastal plain. In these supposed "safe havens," retreat has resulted not from gradual deterioration of environmental conditions, but from abrupt changes wrought by highly questionable private and public policy decisions driven by the short-term interests of local or regional corporate stakeholders. Notable representative examples include the Mississippi River–Gulf Outlet Canal (MRGO) and various deepwater shipping channels connecting interior industrial "ports" directly with the Gulf. Built primarily between the mid-1960s and 2005, these artificial, highly engineered waterways not only contribute significantly to erosion, but they also now serve as major storm surge conduits transporting unimpeded floodwaters directly from the sea to the industrial communities they service.

Major businesses have begun to recognize the artificial waterways' inherent potential for catastrophic flooding events, which could—and should—have been foreseen. In Houma, for example, several major petroleum service corporations have moved their operations from the waterfront to more secure locations several miles to the north.

The ongoing retreat from the increasingly vulnerable Louisiana coast is causing—and will continue to create for the foreseeable future—major social, cultural, and economic repercussions. The prospect of relocation is particularly distressing to native South Louisianians. As recent decennial federal census reports have made abundantly clear, Louisiana's coastal plain is home to the nation's most sedentary population, and, until recent decades, it was not unusual for ninth- or tenth-generation French Louisianians to reside within thirty miles of their family's colonial homestead.

Such deep roots have engendered a sense of place little understood by mainstream Americans accustomed to extreme mobility.

A representative individual reportedly felt alienated in an adjacent community with the same demographic and physical characteristics. Hence a move of even six to ten miles is enough to cause severe separation anxiety as the fabric of the region's still viable extended family structure becomes frayed. Indeed, anecdotal information gleaned from local oral history projects indicates that disputes over a move of less than ten miles have led to divorce. Indeed, the regional population is so firmly rooted that, for some, relocation even within the same neighborhood is unthinkable. In one case, an aged marsh dweller refused his son's request to move less than 200 feet to make it easier for the family to provide daily home care. Peering through his window, the recalcitrant senior had gazed for decades upon a neighboring ancient oak tree, and he could not bear the thought of abandoning that treasured prospect.

The demographic shifts also foreshadow the unraveling of cultural traditions that have been more than two centuries in the making. Relocation of displaced populations inevitably entails rapid cultural change and often loss of identity. This is equally

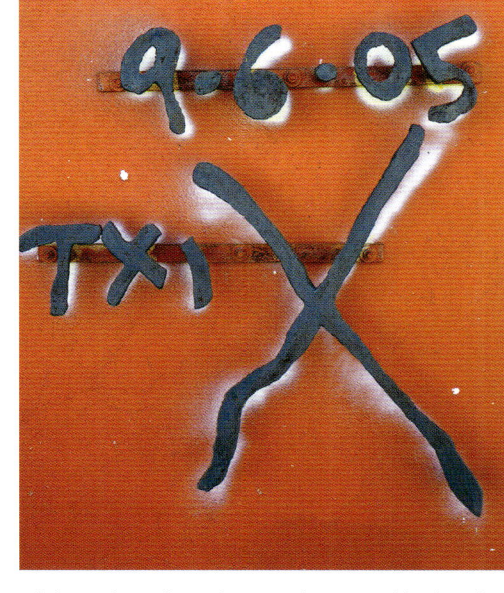

While combing through New Orleans neighborhoods looking for flood victims, searchers identified investigated houses with iconic "X marks," indicating (clockwise) the date on which the structure was inspected, the hazards discovered in the structure, the number of corpses found, and the investigators' individual identification codes. In 2015 some New Orleanians have used the mark as a Roman numeral to indicate resiliency on the tenth anniversary of Hurricane Katrina's passing. (Photo by the authors, 2011)

Signs like this clearly reflected widespread frustration with post-Katrina restoration efforts. (Photo by the authors, 2014)

Throughout Louisiana's coastal zone, the resident population understands the value of the region's disappearing wetlands. (Photo by the authors, 2014)

Logistical companies—often small, family-owned firms—provide a wide variety of essential services, including legal counsel, wireline technologies, hardware manufacturing and replacement, well-log libraries, towing, waste disposal, food distribution, surveying, safety monitoring, oil-spill mitigation, transportation (by air and sea), air-conditioning and refrigeration, groceries, boat charters and rentals, office supplies, laundry, health care, and the list continues.

Hurricane Rita (September 24, 2005) destroyed most of the Louisiana coastline spared by Hurricane Katrina the previous month. In Henry, approximately ten miles inland and seventy-seven linear miles from the storm's landfall in Calcasieu Parish, the Catholic church was inundated and severely damaged by the storm surge. (Photo by the authors, 2005)

true on Lake Pontchartrain's northern shore and in the large communities of Hurricane Katrina and Hurricane Rita ex-pats in Baton Rouge and Lafayette (where they initially styled themselves the "Krewe of Katrina"). Governmental plans to relocate additional population groups—particularly Native Americans—from the coastal plain to urban centers will unquestionably result in wholesale cultural extinction within two generations, despite contrarians' claims of relic population survivals in urban settings.

Demographic shifts pose three major challenges to the state and nation. First, wetlands dwellers constitute the Gulf Coast's most effective environmental early warning system. Armed with environmental knowledge accumulated over more than two centuries and complemented by daily observations of environmental conditions, trappers, oystermen, shrimpers, recreational guides, and small aircraft pilots have traditionally been the first to observe critical changes to the regional ecosystem, including the emergence of dead zones; oil spill impacts; onset of hurricane-driven erosion; climatically induced changes to the Mississippi Flyway; subsidence; changes in the freshwater-saltwater boundary; changes in mammalian, avian, and reptilian habitats; variations in regional hydrology; saltwater intrusion; and alterations in shorelines. Second, the region is the nation's largest single source of oysters and shrimp and is also an important supplier of menhaden, which is used in the production of numerous industrial products ranging from aquaculture feedstock to cat food, to cosmetics, to Omega 3 tablets. Finally, the ongoing human outmigration poses a genuine threat to America's energy security. The nation's vital Gulf offshore oil industry is completely dependent upon the availability of a highly skilled, resident coastal plain workforce to staff seismographic operations; operate the onshore and offshore exploration and production platforms and rigs; construct and maintain pipelines; build and operate the flotilla of boats and ships needed by the industry; provide diving, helicopter, and aircraft services; and supply a vast array of other vital logistical energy industry services.

Depletion of the local workforce threatens the industry's current vitality and jeopardizes its ability to capitalize upon future developmental opportunities. This is particularly true of deep-water drilling operations, ultra-deep onshore wells, and the great potential of the area's emerging liquefied natural gas (LNG) infrastructure that is rapidly morphing from intended import facilities to operational export installations servicing global natural gas markets.

As the domestic shale oil and gas industry continues to boom, South Louisiana's LNG facilities are a noteworthy part of a new "energy revolution." This activity has brought about extraordinary recent manufacturing and industrial growth in South Louisiana. The ability to export liquefied natural gas is essential to the health of the US energy sector, the nation's security, and, in the wake of Russia's recent (2015) use of petroleum exports for political coercion, perhaps the economic independence of western Europe as well.

Existing circumstances, however, jeopardize the human infrastructure required to sustain such enterprises. In the wake of the benchmark hurricanes of 2005, the Federal Emergency Management Agency (FEMA) and insurance companies have promulgated often conflicting regulations that make continued occupation of the coastal wetlands extremely difficult, if not impossible. For example, FEMA has mandated the elevation of homes in the coastal plain as a prerequisite for national flood insurance, and the agency has provided subsidies to homeowners to make the lifting possible. In some areas, FEMA regulations require elevation of residences more than eleven feet. However, some private insurance companies have refused to provide homeowner's coverage to residents in compliance with FEMA guidelines because the elevated structures are more vulnerable to wind damage.

Such bureaucratic opéras bouffes will undoubtedly profoundly shape the destiny of Louisiana's coastal region. Without homeowners' insurance, residents without independent means will not be able to secure bank mortgages for new construction and/or posthurricane reconstruction. Eventually, residential patterns will consequently exist exclusively within the safe confines of protection levees.

Gone are the days when the course of the coastal region's development was dictated by environmental and economic factors. Instead, the course of history now flows through channels whose control has been co-opted by bureaucrats who often have little firsthand knowledge of the areas impacted by their decisions.

The emergence of bureaucratic entities—both corporate and governmental—as the final arbiters of demography has radically altered the trajectory of human history in the Louisiana coastal plain. For more than 200 years, resilient and fiercely independent coastal dwellers have carved their respective niches in the marshlands, taking up residence wherever economic opportunities and personal whim dictated. Now that this organic decision-making process has been effectively preempted by external agencies, often without meaningful input from long-established residents, it is painfully clear that Louisiana's coastal plain has started a new historical chapter.

For the native population, this new era is literally and figuratively fraught with danger. Residents watch with horror as their comfort zone disappears and the world they have known seemingly turns on its head. Change is inevitable, and, indeed,

To help rehabilitate America's coastal wetlands, local and state governments must fund, maintain, and operate storm-surge barriers, floodgates, levees, and associated structures. Such systems cannot exist without a permanent source of annual funding, In Louisiana, all of these elements constitute vital parts of the state's last line of defense. Although federal funding is crucial to the survival of the wetlands, the state should not rely too heavily on such capricious external sources of revenue.

coastal dwellers for generations have had to adjust to continuously changing environmental conditions, waxing and waning economic markets, and fluctuating wildlife populations. However, even in the worst of times, the coastal wetlands always provided its human denizens a safety net: food, shelter, and jobs, as well as a sense of mental well-being afforded by the cultural cocoon preserved by the solitude of the marshes. Now, the marshes' living verdant carpets, the wetlands' wildlife, and the way of life built around them are in dire peril—if the past is indeed a prelude. Dozens of nineteenth-century communities—including St. Malo, Manila Village, Bassa-Bassa, Alluvial City, Coon Road, Daisy, Chong Song, Cabinish, Camp Dewey, Dunbar, Falia, Balize, Avoca Island, Nichols, Ostrica, Seabreeze, Doullet's Canal, Oysterville, Perry, English Lookout, Fisherman's Village, Chênière Caminada, Yankee Camp, and many others—now exist only as vestigial memories in the region's historical literature, for their respective sites now lie beneath the waves of encroaching Gulf waters or as clusters of aging pilings. The economic infrastructure that once sustained these lost settlements has also vanished.

The geological and economic changes that claimed these marsh outposts, now complemented by intrusive bureaucratic regulations, collectively assail the surviving marshland communities, which calls into question the prospect of their long-term survival. Are individuals, corporations, and governments willing to exhibit the initiative, leadership, vision, and courage necessary to meet unprecedented challenges and place the common good above narrow self-interests? Are they also prepared to commit the financial, physical, and human resources necessary to literally turn the tide and rescue a precious national resource? Or will these beleaguered coastal communities join the ranks of coastal Louisiana businesses, nonrenewable resources, species, and communities that, in local parlance, "ain't there no more"?

Louisiana's coastal dwellers have kept alive the memory of camps as primitive, spartan working installations, and, as a consequence, "poverty" remains a dominant subtext in local camp naming conventions. The role of modern camps, however, has clearly evolved. Most are now focal points for recreational family outings. (Photo by the authors, 1997)

Coastal erosion in Louisiana has claimed a surface area approximately the size of Delaware since 1930. Efforts to combat land loss will be extremely expensive, but the alternative will be still more costly. (Photo by the authors, 1983)

Flooding

The 1927 flood, dubbed America's worst peacetime disaster by contemporary national media, inundated an area approximately equivalent to the surface area of Illinois. In lower Louisiana's wetlands, floodwaters displaced approximately 100,000 residents from the western fringe of the Atchafalaya Basin alone. The public outcry for a "permanent" solution to the perennial threat posed by effectively unfettered waterways resulted in construction of levee and flood-control systems of unprecedented scale that diverted life-giving nutrients away from the coastal wetlands. (From the authors' collections)

An aerial view of the town of Indian Village (eastern edge of the Atchafalaya Basin, Iberville Parish), inundated by floodwaters from the 1927 flood event. (Photo courtesy of the National Archives and Records Administration, Prints & Photographs Division, College Park, MD, photographer unknown, 1927, call number RG 77-MRC, Box 1 - 3-2)

Refugees stranded on a levee by rapidly rising floodwaters, 1927. (Photo courtesy of the National Archives and Records Administration, Prints & Photographs Division, College Park, MD, photographer unknown, 1927, call number RG 77-MRC, Box 1 - 5-1)

The US Coast Guard, employing speedboats formerly used against rumrunners on the East Coast, rescued a large but as yet underdetermined number of lower Louisiana residents trapped by floodwaters on their roofs and along levees. (Photo courtesy of the National Archives and Records Administration, Prints & Photographs Division, College Park, MD, photographer unknown, 1927, call number RG 77-MRC, Box 1 - 3-2)

The steamer *Tuscumbia* discharging "refugees, chickens, stock and household articles." The refugees then boarded a barge provided by the Louisiana Navigation and Railway Company for transportation to refugees' camps. (Photo courtesy of the National Archives and Records Administration, Prints & Photographs Division, College Park, MD, photographer unknown, 1927, call number, RG 77-MRC, Box 1 - 3-2)

In May 2011 the Atchafalaya Basin experienced near-record water levels. The peripheral levees protected adjoining residential areas from flooding but also diverted most of the floodwaters' invaluable sediments away from eroding coastal wetlands. (Photo by the authors, 2011)

In Louisiana, the 1927 flood literally washed away several rural communities—not only homes, barns, ancillary buildings, and fences, as depicted here, but also, in many cases, the very topsoil that made agriculture possible on the farmers' once verdant fields. (Photo courtesy of the Center for Louisiana Studies, University of Louisiana at Lafayette, photographer unknown, 1927)

On September 24, 2005, Hurricane Rita, the fourth most intense Atlantic hurricane ever recorded, made landfall near the Louisiana-Texas border. Rita's record storm surge irreparably damaged thousands of homes in the coastal plain. (Photo courtesy of the Center for Louisiana Studies, University of Louisiana at Lafayette, 2005)

Most of Louisiana's 1927 flood refugees initially made their way to tent cities maintained by the American Red Cross, like this one at Lafayette. In these camps, these overwhelmingly rural folk received their first exposure to such amenities as indoor plumbing and electricity, which were becoming commonplace in urban areas. (Photo courtesy of the Center for Louisiana Studies, University of Louisiana at Lafayette, photographer unknown, 1927)

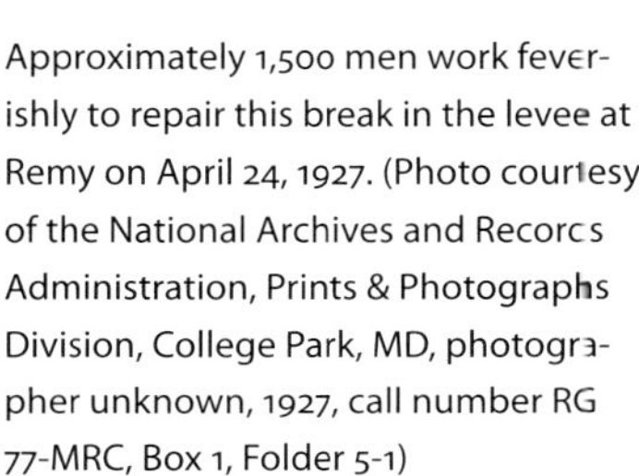

Approximately 1,500 men work feverishly to repair this break in the levee at Remy on April 24, 1927. (Photo courtesy of the National Archives and Records Administration, Prints & Photographs Division, College Park, MD, photographer unknown, 1927, call number RG 77-MRC, Box 1, Folder 5-1)

At the height of the 1927 flood, New Orleans political power brokers successfully lobbied the federal government to dynamite the levees at Caernarvon, in St. Bernard Parish, in order to relieve pressure on New Orleans levees and thus spare the city from inundation. The floodwaters released by the resulting artificial crevasse flooded St. Bernard Parish and displaced its approximately 10,000 residents. Modern studies of the flood have concluded that the Caernarvon crevasse was unnecessary, as crevasses then developing upstream were naturally lowering water levels in the swollen Mississippi. (Photo courtesy of the National Archives and Records Administration, Prints & Photographs Division, College Park, MD, photographer unknown, 1927, call number RG 77-MRC, Box 2, image Caernarvon, 7866-11a)

Hurricanes

This scene is immediately recognizable to hurricane and flood victims in the coastal plain. Following Hurricane Rita, residents of rural Vermilion Parish—many of whose homes were inundated to the eaves—were compelled to completely gut the interiors of their residences to mitigate the health hazard of toxic black mold. In the Henry area, pictured here, some inhabitants were compelled to repeat this gut-wrenching process three times in a five-year span. (Photo by the authors, 2005)

In the wake of Hurricane Rita, lower Vermilion Parish ranchers were compelled to move their surviving herds to higher ground as quickly as possible. Cattle that drank even brackish water suffered terminal liver damage. (Photo courtesy of the Center for Louisiana Studies, University of Louisiana at Lafayette, 2005)

October 30, 1886.] FRANK LESLIE'S ILLUSTRATED NEWSPAPER. 169

LOUISIANA.—THE RECENT INUNDATION AND TERRIBLE LOSS OF LIFE AT JOHNSON'S BAYOU—RELIEF PARTIES SEARCHING FOR VICTIMS OF THE STORM.

From a Sketch by a Corresponding Artist.—See Page 166.

This lithograph shows the destruction wrought in the Johnson Bayou area (Cameron Parish) by the terrible 1886 hurricane, which claimed a significant but still undetermined number of victims. (Lithograph from *Frank Leslie's Illustrated Newspaper*, October 30, 1886, 169)

Foundations were literally all that remained of the Holly Beach resort community following Hurricane Rita's passage in September 2005. Once derisively known as the Cajun Riviera for its modest, often poorly maintained camps, it has recently been reborn as an upscale beach community. (Photo by the authors, 2005)

After the hurricane seasons of 2005 and 2008, some homeowners in the small beachfront community of Holly Beach announced, in hand-painted signs, their willingness to sell their property. (Photo by the authors, 2011)

Johnson Bayou, a small unincorporated community east of the Sabine River, was extensively damaged by Hurricane Ike in 2008. The reopening of a school, post office, and grocery store is a sign of a community's return. (Photo by the authors, 2010)

Hurricanes are quick, deadly, and traumatizing events and become mental benchmarks in the affected populations' individual and collective psyche. After 2005, Louisiana was devastated by a series of hurricanes; however, before storms were named, the year of the storm became the memory point, such as 1886, 1893, 1909, 1915, and 1947. After 1947, hurricanes named Flossy, Audrey, Ethel, Carla, Hilda, Betsy, and Camille are locked permanently in the region's collective memory. This May 1969 photo captures the destructive force of Hurricane Camille to a Louisiana Wildlife and Fisheries camp at Garden Island Bay in the Mississippi delta. (Photo courtesy of the Louisiana Department of Wildlife and Fisheries)

Recovery and Protection

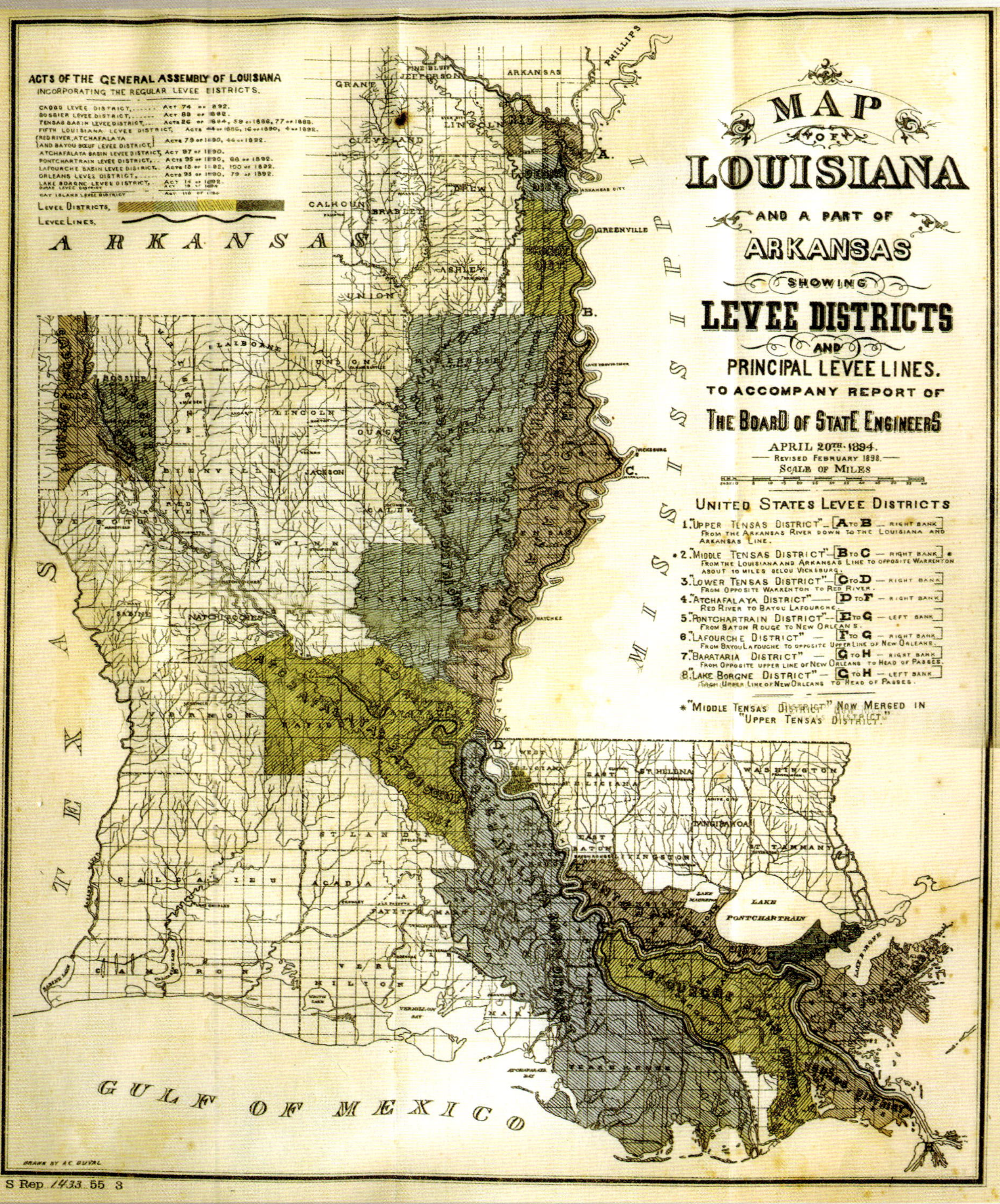

Since the post–Civil War era, Louisiana has had a balkanized system of levee control, with administrative responsibility lodged in the hands of highly localized boards. This 1894 map shows the levee districts as they then existed in Louisiana. (From the authors' collections)

The Bonnet Carré spillway control structure during the 2011 flood. (Photo by the authors, 2011)

As part of the flood-control master plan implemented after the 1927 flood, the US Army Corps of Engineers established spillways designed to channel floodwaters away from populated areas. This image depicts land-clearing operations preceding construction of the Bonnet Carré Spillway's upper levee in September 1929. (Photo courtesy of the National Archives and Records Administration, Prints & Photographs Division, College Park, MD, photographer unknown, 1927, call number RG 77-MRC, Box 2, Bonnet Carré)

The Terrebonne Parish government's concern about coastal erosion led to an effort to use sand fencing in an effort to slow sand migration. This technique, used in the 1970s and early 1980s, constituted one of the earliest attempts to stabilize Louisiana's barrier islands. (Photo by the authors, 2004)

The twin coastal disasters of 2005 (Hurricanes Katrina and Rita) drove home to coastal plain residents the importance of the marshes as storm buffers. Grass-roots pressure has compelled the state and federal governments to undertake some modest reclamation efforts, such as the one pictured here in lower Vermilion Parish. However, the largely experimental engineering projects to this point have been quite insufficient to reverse disastrous historical trends in coastal erosion. (Photo by the authors, 2007)

This image documents one of the earliest recorded efforts to transplant grass to combat erosion on Marsh Island. (Photo courtesy of the Louisiana Department of Wildlife and Fisheries)

After the four storms associated with the 2005 and 2008 hurricane seasons, there was a conscious effort, often driven by unprecedented insurance premiums, to elevate homes in the communities devastated by one or more of these storms. (Photo by the authors, 2006)

After Hurricane Katrina in 2005, the US Army Corps of Engineers spent more than $14 billion fortifying New Orleans against future storm surge events. This is the first phase of what could eventually be a $20 billion effort to raise the city's protection levees. The ongoing threat of inundation consists of three main factors: sea-level rise, subsidence, and coastal land loss. Consequently, while it can be argued that New Orleans is safer thanks to the new levees, it is not safe. As Hurricane Sandy demonstrated, other coastal cities, such as New York, Miami, Baltimore, Tampa, and Charleston, are equally vulnerable to inundation. Such urban centers cannot be abandoned, and the nation will have to come to grips with this increasingly intractable—and expensive—problem. (Photo by the authors, 2011)

Louisiana finds itself at a crossroads, and only time will tell if the sun is rising or setting on a critical era in the coastal plain's storied history. (Photo by the authors, 2014)

AFTERWORD

Disgraced former Speaker of the US House of Representatives Dennis Hastert's now notorious comments about the wisdom of rebuilding post-Katrina New Orleans, which, in his judgment, should have been largely bulldozed, ripped the veil off a festering double standard in American public attitudes toward southern coastal communities in general and Gulf Coast municipalities in particular. Before Hurricane Katrina, public policy polling commissioned by the Louisiana state government very clearly showed that residents of the American heartland had no interest whatsoever in government subsidies for wetland reclamation projects that ultimately would have protected the Crescent City from the catastrophic events of August 2005. Instead, prevailing popular attitudes suggested that coastal residents should abandon their homes because of the persistent threat posed by seasonal hurricanes and tropical storms. Yet no contemporary poll respondent—much less a national politician—would have dared to suggest that persons in earthquake zones or tornado alleys should relocate, despite the acknowledged risks.

The massive destruction wrought by Hurricane Sandy (2012) along the northeastern Atlantic coastline has initiated a seismic shift in national attitudes. Prior to Sandy, the coastal population above the Mid-Atlantic states enjoyed a false sense of security born of a pervasive sense of hurricane immunity. The Category 1 storm, however, clearly demonstrated that, as a result of the climatological volatility born of rapid climate change, no coastal community is safe from the onslaught of tropical weather systems.

The shattered sense of invulnerability has forced communities impacted by Sandy to reassess their abilities to not only withstand such natural disasters but also to ensure that devastated communities rebound quickly and systematically. In reconstructing their shattered world and preparing for future meteorological challengers, northeastern political and business leaders have come to realize that critical community resiliency is entirely dependent upon proactive measures tailored to, and meaningfully shaped by, the at-risk population.

Louisiana's government, which has traditionally utilized a top-down approach to coastal affairs, should perhaps profit from its northern counterparts' example by thoroughly reexamining the role of the most directly impacted populations in decision-making processes. As the Jersey Shore has discovered, communities will survive only if existing residents are encouraged to maintain their attachment to their land and restore the physical space where they live, work, and play.

Effective communication with the Bayou State's at-risk coastal populations will require an overhaul of bureaucratic attitudes and approaches, because administrators have traditionally viewed Louisiana's coastal population as either a persistently cantankerous annoyance or as an exotic, mysterious, unapproachable, and vaguely sinister enclave. Establishment of effective bilateral communications between the two groups may require an approach perfected by America's advertising industry in the 1950s and 1960s. During the Golden Age of Television, Madison Avenue firms discovered their clients could greatly expand their respective market shares by more effectively approaching a broad spectrum of racial and ethnic groups. Louisiana should embrace this model to ensure that the coastal population is more directly involved in restoration/rehabilitation efforts. Established communities are living repositories of traditional ecological knowledge capable of providing insights into the subtleties of coastal landscapes essential to the success of many, perhaps all, restoration efforts. In addition, policy makers must also recognize and come to terms with the immutable political realities of the coast's unfolding environmental crisis: volumetric water flow rates, sediment load, subsidence, sea level rise, climate change, and at-risk flora and fauna are all important issues that must be part of the restoration dialogue. However, the bottom line is that problems cast no ballots, and local voters—the seemingly invisible endangered species—will ultimately mandate solutions.

Communities that lost their post offices and may have been abandoned

Parish	Communities
Cameron	Berry, Grand Lake, Lakeside, Leesburg, Premo, Radford, Shell Bank
Iberia	Belle Place, Burke Station, Duboin, Grand Cote, Hubertville, Murphy, Oasis, Patoutville
Jefferson	Amesville, Belgrove, Camindaville, Camp Parapet, Carrollton, Clarkville, Geraty, Jefferson, McDonoghville, Rice, Waggaman, Walker Town
Lafourche	Belle Amie, Bowie, Florence, Fort Guion, Guidry's, Harang, Interior Parish, La Croix, Lafourche Crossing, Lidivine, Malagay, Norah, Orange City, Pugh, Rathborne, Sassone, Sepine's, Thibodeauxville, Toup's
Orleans	Algiers, Chef, Chef Menteur, Coquille, Fort Pike, Lee, Little Woods, Petites Coquille, Pontchartrain Grove, Rigolets, South Point, Southpoint, Spanish Fort
Plaquemine	Adema, Balize, Beausejour, Bel Air, Belair, Benjamin, Bohemia, Buras Settlement, City Price, Concession, Dime, Duvic, English Turn, Fort Jackson, Grand Folly, Grand Prairie, Home Place, Jesuit's Bend, Jump, Junior, Lawrence, Martin, Moss Grove, Naomi, Neptune, Nero, Nestor, Nicholls, Orillon's Farm, Pigniolo, Pilot Town, Point Michael, Quarantine, Red Store, Saint Sophie, South West Pass, Sunrise, The Forts, Vandyke, Wood Park
St. Bernard	Alluvial, Bienvenu, Drew's Station, Ducros Station, Estopinal, Hopedale, La Chinche, Melonie, Poydras, Shell Beach, Terre Aux Boeufs, Verret
St. Mary	Acklen, Alligator, Bartels, Bayou Ramois, Belle Isle, Calumet, Carlin Settlement, Centreville, Coteau, Crawford, Cypremort, Darrall, Dutch Settlement, Glencoe, Glenwild, Grand Woods, Irish Bend, La Teche, Maillard, North Bend, Pattersonville, Pharr, Ramos, Saint Philip, Scaly, Teche, Winsted, Wooster
Terrebonne	Ardoyne, Belle Grove, Cazeaux, Daspit, Donner, Falgout, Gibson City, Jewell, Lacache, Laurence, Live Oak, Ormond, Point Farm, Saint Eloi, Terre Bonne, Tigerville, Williamsburgh
Vermilion	Adlar, Choate, Cossinade, Cow Island, Dixie, Florence, Grand Marais, Gregg, Hartman, Laurents, Leleux, Millington, Neville, Parc, Pardue, Peigneur, Perry's Bridge, Ramsey, Riceville, Shell Beach, Theal, Tortue, Wester Ogle

APPENDIX A

Post Office Closures

APPENDIX B

The Rice Harvest in Louisiana, 1851

Parish	Pounds of Rice
Ascension	33,500
Assumption	99,770
Avoyelles	291,350
E. Baton Rouge	4,009
W. Baton Rouge	900
Bienville	6,688
Bossier	145
Caddo	10
Calcasieu	1,176
Caldwell	2,820
Carroll	No Return
Catahoula	206
Claiborne	60
Concordia	100
De Soto	88,400
E. Feliciana	42,675
W. Feliciana	8,000
Franklin	76
Iberia	No Return
Jackson	5,070
Jefferson	122,000
Lafayette	2,168
Lafourche	281,989
Livingston	83,480
Madison	8,150
Morehouse	17,235
Natchitoches	14,375
Orleans	40,000
Ouachita	420
Plaquemines	1,586,740
Pointe Coupée	16,840
Rapides	4,500
Sabine	21,180
St. Bernard	No Return
St. Charles	619,000
St. Helena	54,868
St. James	68,500
St. John the Baptist	814,200
St. Martin	8,700
St. Mary	140
St. Tammany	97,793
Tensas	3,000
Terrebonne	466,900

Parish	Pounds of Rice
Union	No Return
Vermilion	1,664
Washington	150,750

Source: J. Blodget Britton, "Rice, Its Culture in Louisiana," *Thibodaux Minerva* (June 17, 1851), 1.

BIBLIOGRAPHY

Primary Sources

Articles

"Child Labor and the Work of Mothers in Oyster and Shrimp Canning Communities on the Gulf Coast." Bureau Publication, No. 98. Washington, DC, 1922. In *Suffer the Little Children: Two Children's Bureau Bulletins*, ed. L. Stein. New York: Arno Press, 1977.

Johnson, C. "Life in the Louisiana Swamp." *Outing Magazine* 45 (1905): 515–526.

"Terrebonne Parish." *Commercial Review of the South and West*, New Series, 2 (1850): 147.

Books, Monographs, and Pamphlets

Brasseaux, Carl A., and Jacqueline K. Voorhies, eds. *Quest for the Promised Land: Official Correspondence Relating to the First Acadian Migration to Louisiana, 1764–1769*. Trans. Carl A. Brasseaux, Emilio Fabian Garcia, and Jacqueline K. Voorhies. Lafayette: Center for Louisiana Studies, 1989.

Clark, E. D., L. McNaughton, and M. E. Pennington. *Shrimp: Handling, Transportation and Uses*. Washington, DC: US Department of Agriculture, 1917.

Darby, William. *The Emigrant's Guide to the Western and Southwestern States and Territories*. New York: Kirk and Mercein, 1818.

———. *A Geographical Description of the State of Louisiana*. Philadelphia: John Melish, 1816.

Dennett, Daniel. *Louisiana As It Is*. New Orleans: Eureka Press, 1876.

Dunbar, G. W. *Original Dunbar Shrimp*. New Orleans: Dunbar-Dukate, 1915.

Ellet, Charles, Jr. *The Mississippi and Ohio Rivers*. Philadelphia: Lippincott, Grambo, 1853.

Flint, Timothy. *A Condensed Geography and History of the Western States, or The Mississippi Valley*. 2 vols. Cincinnati: E. H. Flint, 1828.

Goode, G. B. *The Fisheries and Fishery Industries of the United States*. Section V, Vol. 1. Washington, DC: Government Printing Office, 1887.

Hallowell, C. *People of the Bayou: Cajun Life in Lost America*. New York: E. P. Dutton, 1979.

Humphreys, A. A., and H. L. Abbott. *Report Upon the Physics and Hydraulics of the Mississippi River; Upon the Protection of the Alluvial Region Against Overflow; and Upon the Deepening of the Mouths: Based Upon Surveys and Investigations Made Under the Acts of Congress Directing the Topographical and Hydrographical Survey of the Delta of the Mississippi River, with Such Investigations as Might Lead to Determine the Most Practicable Plan for Securing It from Inundation, and the Best Mode of Deepening the Channels at the Mouths of the River*. Submitted to the Bureau of Topographical Engineers, War Department, 1861 (No. 4). Washington, DC: J. B. Lippincott, 1861.

Hutchins, Thomas. *An Historical Narrative and Topographical Description of Louisiana and West-Florida*. Facsimile reprint. Gainesville: University of Florida Press, 1968.

Kemper, J. P. *Floods in the Valley of the Mississippi: A National Calamity, What Should Be Done about It*. New Orleans: National Flood Commission, 1928.

Louisiana Bureau of Statistics of Labor. *Ninth Biennial Report of the Department of Commissioners of Labor and Industrial Statistics of the State of Louisiana*. New Orleans: Hauser Printing, 1918.

Louisiana Oyster Commission. *First Annual Report of the Oyster Commission of Louisiana*. New Orleans: American Printing, 1904.

National Canners Association. *Canners Directory*. Washington, DC: National Canners Association, 1922.

Tompkins, Frank H. *Riparian Lands of the Mississippi River: Past—Present—Prospective*. New Orleans: privately printed, 1901.

United States Congress. *American State Papers: Documents, Legislative and Executive of the Congress of the United States, in Relation to the Public Lands*. Vol. 5. Washington, DC: Duff Green, 1834.

United States Department of Commerce, Bureau of the Census. *Manufactures, 1905*. Vol. 2. Washington, DC: Government Printing Office, 1907.

———. *Thirteenth Census of the United States Taken in the Year 1910*. Vol. 9, *Manufactures, 1909*. Washington, DC: Government Printing Office, 1913.

Wyckoff, William C. *Report on the Silk Manufacturing Industry of the United States*. Washington, DC: Government Printing Office, 1883.

Manuscripts

Royal Memoir to Serve as Instructions for *Ordonnateur* Edmé Salmon, May 15, 1731. Archives Nationales, Paris, Archives des Colonies, Series B, Vol. 55, Folio 611vo.

Newspapers

Biloxi Daily Herald, 1902.

Secondary Sources

Articles

Adkins, G. "Shrimp with a Chinese Flavor." *Louisiana Conservationist* 25 (1973): 20–25.

Allen, C. M., M. Vidrine, B. Borsari, and L. Allain. "Vascular Flora of the Cajun Prairie of Southwestern Louisiana." In *Proceedings of the Seventeenth North American Prairie Conference, Seeds for the Future—Roots of the Past*, ed. N. P. Bernstein and L. J. Ostrander, 35–41. North Mason City: Iowa Area Community College, 2001.

Barton, D. C., and R. H. Goodrich. "The Jennings Oil Field, Acadia Parish, Louisiana." *Bulletin of the American Association of Petroleum Geologists* 10 (1926): 72–92.

Baumann R. H., J. W. Day, and C. A. Miller. "Mississippi Deltaic Wetland Survival: Sedimentation versus Coastal Submergence." *Science* 224 (1984): 1093–1095.

Bell, H. W., and R. A. Cattell. "The Monroe Gas Field: Ouachita, Morehouse and Union Parishes, Louisiana." *Louisiana Department of Conservation Bulletin* 9 (1921).

Bilbo, W. R. "Construction Is Different in Marsh-Water Areas." *Pipeline Engineer* 29 (1957): D-25–D-28.

Bearss, Edwin C. "The Civil War Comes to the Lafourche." *Louisiana Studies* 5 (1966): 97–155.

Bunya, S., J. C. Dietrich, J. J. Westerink, L. H. Holthuijsen, C. Dawson, R. A. Luettich Jr., R. Jensen, J. M. Smith, G. S. Stelling, and G. W. Stone. "A High Resolution Coupled Riverine Flow, Tide, Wind, Wind Wave, and Storm Surge Model for Southern Louisiana and Mississippi. Part I: Model Development and Validation." *Monthly Weather Review* 138 (2010): 345–377.

Byrnes, M. R., R. A. McBride, Q. Tao, and L. Duvic. "Historical Shoreline Dynamics along the Chenier Plain of Southwestern Louisiana." *Transactions of the Gulf Coast Association of Geological Societies* 45 (1995): 113–122.

"Caddo Lake Holds Distinction of Being the Nation's First Marine-Drilling Operation." *Oil and Gas Journal* 49 (1950): 267–272.

Coleman, J. M. "Dynamic Changes and Processes in the Mississippi River Delta." *Bulletin of the Geological Society of America* 100 (July 1988): 999–101.

Coleman, J. M., H. H. Roberts, and G. W. Stone. "Mississippi River Delta: An Overview." *Journal of Coastal Research* 14 (1993): 698–716.

Comeaux, Malcolm. "The Atchafalaya River Raft." *Louisiana Studies* 9 (1970): 217–227.

Conner, W. H., and M. Brody. "Rising Water Levels and the Future of Southeastern Louisiana Swamp Forests." *Estuaries* 12 (1989): 318–323.

Dauenhauer, J. B., Jr. "Shrimp: An Important Industry of Jefferson Parish." *Jefferson Parish Yearly Review* (1938): 109–131

Davis, Donald W. "Canals and the Southern Louisiana Landscape." In *Geographical Snapshots of North America*, ed. D. G. Janelle, 375–379. New York: Guilford Press, 1992.

———. "Historical Perspective on Crevasses, Levees, and the Mississippi River." In *Transforming New Orleans and Its Environments*, ed. C. E. Colton, 84–106. Pittsburgh: University of Pittsburgh Press, 2000.

———. "Logging Canals: A Distinct Pattern of the Swamp Landscape in South Louisiana." *Forest and People* 25 (1975): 14–17, 33–35.

Day, J. W., D. F. Boesch, E. J. Clairain, G. P. Kemp, and S. B. Laska. "Restoration of the Mississippi Delta: Lessons from Hurricanes Katrina and Rita." *Science* 315 (2007): 1679–1684.

Draut, A. E., G. C. Kineke, O. K. Huh, J. M. Grymes III, K. A. Westphal, and C. C. Moeller. "Coastal Mudflat Accretion under Energetic Conditions, Louisiana Chenier-Plain Coast, USA." *Marine Geology* 214 (2005): 27–47.

Ferleger, Louis. "The Problem of 'Labor' in the Post-Reconstruction Louisiana Sugar Industry." *Agricultural History* 72 (1998): 140–158.

Ginn, Mildred Kelly. "A History of Rice Production in Louisiana to 1896." *Louisiana Historical Quarterly* 23 (1940): 544–588.

Gould, H. R., and E. McFarland Jr. "Geological History of the Chenier Plain, Southwestern Louisiana." *Transactions of the Gulf Coast Association of Geological Societies* 9 (1959): 261–270.

Graves, R. A. "Louisiana, Land of Perpetual Romance." *National Geographic Magazine* 57 (1930): 393–482.

Halpern, Rick. "Solving the 'Labour Problem': Race, Work and the State in the Sugar Industries of Louisiana and Natal, 1870–1910." *Journal of Southern African Studies* 30(1), Special Issue: Race and Class in South Africa and the United States (March 2004): 19–40.

Henry, A. J. "The Floods of 1927 in the Mississippi Basin." *Science*, New Series, 67 (1928): 15–16.

Howe, H. V., R. J. Russell, and J. H. McGuirt. "Physiography of Coastal Southwest Louisiana." In *Reports on the Geology of Cameron and Vermilion Parishes*, ed. R. J. Russell, H. V. Howe, J. H. McGuirt, C. R. Dohm, W. Hadley Jr., F. B. Kniffen, and C. A. Brown, 1–72. Geological Bulletin No. 6. New Orleans: Geological Survey, Louisiana Department of Conservation, 1935.

Jones, Bill. *Louisiana Cowboys*. Gretna, LA: Pelican Publishing, 2007.

Kniffen, F. B. "Louisiana House Types." *Annals of the Association of American Geographers* 26 (1936): 179–193.

———. "Preliminary Report on the Indian Mounds and Middens of Plaquemines and St. Bernard Parishes." In *Lower Mississippi River Delta: Report on the Geology of Plaquemines and St. Bernard Parishes*, ed. R. J. Russell, H. V. Howe, J. H. McGuirt, C. R. Dohm, W. Hadley Jr., F. B. Kniffen, and C. A. Brown, 407–422. Geological Bulletin No. 8. New Orleans: Geological Survey, Louisiana Department of Conservation, 1936.

Lippincott, I. "A History of River Improvement." *Journal of Political Economy* 22 (1914): 630–660.

Lopez, John A. "The Environmental History of Human-Induced Impacts to the Lake Pontchartrain Basin in Southeastern Louisiana since European Settlement, 1718–2002." *Journal of Coastal Research*, Special Issue 54 (2009): 1–11.

———. "The Multiple Lines of Defense Strategy to Sustain Coastal Louisiana." *Journal of Coastal Research*, Special Issue 54 (2009): 186–197.

MacKenzie, C. L., Jr. "History of Oystering in the United States and Canada, Featuring the Eighth Greatest Oyster Estuaries." *Marine Fisheries Review* 58 (1996): 1–78.

Mancil, E. "Pullboatin' on the Blind." *Forests and People* 10(4) (1960): 12–16.

McIntire, W. G. "Correlation of Prehistoric Settlements and Delta Development." *Coastal Studies Institute Technical Report No. 5*. Baton Rouge: Louisiana State University Press, 1954.

Mims, Sam. "Louisiana's Administration of Swamp Land Funds." *Louisiana Historical Quarterly* 28 (1945): 277–325.

Neill, C. F., and M. A. Allison. "Subaqueous Deltaic Formation on the Atchafalaya Shelf, Louisiana." *Marine Geology* 214 (2005): 411–430.

Norgress, Rachael Edna. "The History of the Cypress Lumber Industry in Louisiana." *Louisiana Historical Quarterly* 30 (1947): 979–1059.

Orr, A. "Shrimp Canning at Biloxi, Mississippi." *Canning Age* 2(1) (1921): 36–39.

Pabis, George S. "Delaying the Deluge: The Engineering Debate over Flood Control on the Lower Mississippi River, 1846–1861." *Journal of Southern History* 64 (1998): 421–454.

Padgett, H. R. "Physical and Cultural Associations of the Louisiana Coast." *Annals of the Association of American Geographers* 5(3) (1969): 481–493.

———. "Some Physical and Biological Relationships to the Fisheries of the Louisiana Coast." *Annals of the Association of American Geographers* 56(3) (1966): 423–439.

Pearcy, Matthew T. "A History of the Ransdell-Humphreys Flood Control Act of 1917." *Louisiana History* 41 (200): 133–159.

Penland, S., and K. E. Ramsey. "Relative Sea-Level Rise in Louisiana and the Gulf of Mexico: 1908–1988." *Journal of Coastal Research* 6 (1990): 323–342.

Penland, S., H. F. Roberts, S. J. Williams, A. H. Sallenger Jr., D. R. Cahoon, D. W. Davis, and C. G. Groat. "Coastal Land Loss in Louisiana." *Transactions of the Gulf Coast Association of Geological Societies* 40 (1990): 685–700.

Perrin, Warren. *Vermilion Parish*. Mt. Pleasant, SC: Arcadia Publishing, 2011.

Post, Lauren C. "The Old Cattle Industry of Southwest Louisiana." *McNeese Review* (1957): 43–55.

Postgate, J. C. "History and Development of Swamp and Marsh Drilling Operations." *Oil and Gas Journal* 47(48) (1949): 87.

Roberts, H. H. "Dynamic Changes of the Holocene Mississippi River Delta Plain: The Delta Cycle." *Journal of Coastal Research* 13(3) (1997): 605–627.

Robinson, H. R. "Handling Shrimp in the Canning Plant." *Proceedings of the Gulf and Caribbean Fisheries Institute* 6 (1954): 17–20.

Russell, R. J. "Flotant." *Geographical Review* 32(1) (1942): 74–98.

Russell, R. J., and H. V. Howe. "Cheniers of Southwestern Louisiana." *Geographical Review* 25(3) (1935): 449–461.

Turner, R. E., and M. E. Boyer. "Mississippi River Diversions, Coastal Wetland Restoration/Creation and an Economy of Scale." *Ecological Engineering* 8(2) (1997): 117–128.

Visser, J. M., C. E. Sasser, R. H. Chabreck, and R. G. Linscombe. "Marsh Vegetation Types of the Mississippi River Deltaic Plain." *Estuaries* 21 (1998): 818–828.

Waddill, R. D. "History of the Mississippi River Levees, 1717 to 1944." *Memorandum*. Vicksburg, MS: War Department Corps of Engineers, Mississippi River Commission, 1945.

Waldo, E. "From Cattle to Shrimp Boats." *Louisiana Conservationist* 17(7–8) (1965): 18–21.

Williams, N. "Drilling for Oil in the Out of the Way Marshlands of Terrebonne Parish, Louisiana." *Oil and Gas Journal* 28 (December (1929): 40–41.

Williams, S. J., S. Penland, and A. H. Sallenger, eds. "Louisiana Barrier Island Erosion Study: Atlas of Barrier Shoreline Changes in Louisiana from 1853 to 1989." *Miscellaneous Investigations Series I-2150-A*. Washington, DC: US Geological Survey, 1992.

Willis, Edwin E. "Notes for a History of St. Martin Parish." Typescript, 1957. Jefferson-Caffery Louisiana Room, Dupré Library, University of Louisiana at Lafayette.

"World's Largest Shrimp Cannery." *Louisiana Conservation Review* (Spring 1938): 21–24.

Zacharie, F. C. "The Louisiana Oyster Industry." *Bulletin of the United States Fish Commission* 17 (1897): 297–304.

Books, Monographs, and Pamphlets

Ancelet, Barry Jean, Jay D. Edwards, and Glen Pitre. *Cajun Country*. With additional material by Carl A. Brasseaux et al. Jackson: University Press of Mississippi, 1991.

Austin, D. E., et al. *History of the Offshore Oil and Gas Industry in Southern Louisiana*. Vol. 1, *Papers on the Evolving Offshore Industry*. New Orleans, OCS Study MMS 2008–042, 2008.

Baird, S. F., and G. B. Goode. *The History and Present Condition of the Fishery Industries*. Washington, DC: Government Printing Office, 1981.

Barry, John M. *Rising Tide: The Great Mississippi Flood of 1927 and How It Changed America*. New York: Simon and Schuster, 1997.

Baudier, Roger. *The Catholic Church in Louisiana*. New Orleans: Hyatt, 1939.

Bauer, Craig A. *A Leader among Peers: The Life and Times of Duncan Farrar Kenner*. Lafayette: Center for Louisiana Studies, 1993.

Becnel, Thomas A. *The Barrow Family and the Barataria and Lafourche Canal: The Transportation Revolution in Louisiana, 1829–1925*. Baton Rouge: Louisiana State University Press, 1989.

———. *Labor, Church, and the Sugar Establishment: Louisiana, 1887–1976*. Baton Rouge: Louisiana State University Press, 1980.

———. *Senator Allen Ellender of Louisiana: A Biography*. Baton Rouge: Louisiana State University Press, 1995.

Bergerie, Maurine. *They Tasted Bayou Water: A Brief History of Iberia Parish*. New Orleans: Pelican, 1962.

Bernard, Shane K. *The Cajuns: Americanization of a People*. Jackson: University Press of Mississippi, 2003.

———. *Tabasco: An Illustrated History*. Jackson: University Press of Mississippi, 2007.

Bienvenu, Marcelle, et al. *Stir the Pot: The History of Cajun Cuisine*. New York: Hippocrene Books, 2005.

Blume, Helmut. *The German Coast during the Colonial Era, 1722–1803*. Translated, edited, and annotated by Ellen C. Merrill. Destrehan, LA: German-Acadian Coast Historical and Genealogical Society, 1990.

Boutwell, A. P., and G. Folse. *Terrebonne Parish: A Pictorial History, Then and Now*. Houma, LA: Courier, 1997.

Bradshaw, Jim. *Our Acadiana: A Pictorial History of South Louisiana*. [Lafayette, LA]: Thomson South Louisiana Publishing, 1999.

Brasseaux, Carl A. *Acadiana: Louisiana's Historic Cajun Country*. Photographs by Philip Gould. Baton Rouge: Louisiana State University Press, 2011.

———. *Acadian to Cajun: Transformation of a People, 1803–1877*. Jackson: University Press of Mississippi, 1992.

———. *The Founding of New Acadia: The Beginnings of Acadian Life in Louisiana, 1765–1803*. Baton Rouge: Louisiana State University Press, 1987.

———. *"Scattered to the Wind": Dispersal and Wanderings of the Acadians, 1755–1809*. Lafayette: Center for Louisiana Studies, 1991.

Brasseaux, Carl A., and Keith Fontenot. *Steamboats on Louisiana's Bayous: A History and Directory*. Baton Rouge: Louisiana State University Press, 2004.

Brasseaux, Carl A., Keith P. Fontenot, and Claude F. Oubre. *Creoles of Color in the Bayou Country*. Jackson: University Press of Mississippi, 1994.

Brasseaux, Ryan André. *Cajun Breakdown: The Emergence of an American-Made Music*. New York: Oxford University Press, 2009.

Brasseaux, Ryan André, and Kevin S. Fontenot. *Accordions, Fiddles, Two-Step and Swing: A Cajun Music Reader*. Lafayette: Center for Louisiana Studies, 2006.

Broussard, Bernard. *A History of St. Mary Parish*. Baton Rouge, LA: Claitor's, 1977.

Campanella, Richard. *Bienville's Dilemma: A Historical Geography of New Orleans*. Lafayette: Center for Louisiana Studies, 2008.

———. *Geographies of New Orleans: Urban Fabrics Before the Storm*. Lafayette: Center for Louisiana Studies, 2006.

———. *Time and Place in New Orleans: Past Geographies in the Present Day*. Gretna, LA: Pelican Publishing, 2002.

Case, Gladys Calhoon. *The Bayou Chene Story: A History of the Atchafalaya Basin and Its People*. Detroit: Harlo Press, 1973.

Center for Louisiana Studies. *Green Fields: Two Hundred Years of Louisiana Sugar*. Lafayette: Center for Louisiana Studies, 1980.

Chabreck, R. H. *Vegetation, Water, and Soil Characteristics of the Louisiana Coastal Region*. Agricultural Experiment Station, Bulletin No. 664. Baton Rouge: Louisiana State University Press, 1972.

Chamberlain, J. L. *Influence of Hurricane Audrey on the Coastal Marsh of Southwestern Louisiana*. Technical Report No. 10. Baton Rouge: Coastal Studies Institute, Louisiana State University, 1959.

Coleman, J. M. *Recent Coastal Sedimentation: Central Louisiana Coast*. Technical Report No. 17. Baton Rouge: Coastal Studies Institute, Louisiana State University, 1966.

Coleman, J. M., and W. G. Smith. *Late Recent Rise of Sea Level*. Studies of Quaternary Sea Level, Technical Report No. 20, Part D. Baton Rouge: Coastal Studies Institute, Louisiana State University, 1964.

Colten, Craig. *Perilous Place, Powerful Storms: Hurricane Protection in Coastal Louisiana*. Jackson: University Press of Mississippi, 2009.

———. *Transforming New Orleans and Its Environs: Centuries of Change*. Pittsburgh: University of Pittsburgh Press, 2000.

———. *An Unnatural Metropolis: Wrestling New Orleans from Nature*. Baton Rouge: Louisiana State University Press, 2006.

Comeaux, Malcolm. *Atchafalaya Swamp Life: Settlement and Folk Occupations*. Vol. 2, *Geoscience and Man*. Baton Rouge: Louisiana State University Press, 1972.

La Commission des Avoyelles, with Sue L. Eakin. *Avoyelles Parish: Crossroads of Louisiana. Where All Cultures Meet*. Gretna, LA: Pelican Publishing, 1999.

Conrad, Glenn R., ed. *The Cajuns: Essays on Their History and Culture*. Lafayette: Center for Louisiana Studies, 1978.

———, ed. *New Iberia: Essays on the Town and Its History*. Lafayette: Center for Louisiana Studies, 1979.

———, ed. *Saint-Jean-Baptiste des Allemands: Abstracts of the Civil Records of St. John the Baptist Parish to 1803*. 2nd ed. Lafayette: Center for Louisiana Studies, 1992.

Conrad, Glenn R., and Vaughan Baker, eds. *Louisiana Gothic: Recollections of the 1930s*. Lafayette: Center for Louisiana Studies, 1984.

Conrad, Glenn R., and Carl A. Brasseaux. *Crevasse! The 1927 Flood in Acadiana*. Lafayette: Center for Louisiana Studies, 1994.

Costello, Brian. *Devastation Unmeasured: The Tragic History of Floods in Pointe Coupée Parish, Louisiana*. New Roads, LA: New Roads Printing, 2007.

———. *A History of Pointe Coupée Parish*. [New Roads, LA]: privately printed, 1999.

Cowdrey, Albert E. *Land's End: A History of the New Orleans District, U.S. Army Corps of Engineers, and Its Lifelong Battle with the Lower Mississippi and Other Rivers Wending Their Way to the Sea*. New Orleans: US Army Corps of Engineers, 1977.

Daniel, Pete. *Deep'n as It Come: The 1927 Mississippi River Flood*. New York: Oxford University Press, 1977.

Daspit, Fred. *Louisiana Architecture, 1840–1860*. Lafayette: Center for Louisiana Studies, 2006.

Davis, Donald W. *Washed Away: The Invisible Peoples of Louisiana's Wetlands*. Lafayette: University of Louisiana Press, 2010.

Delfino, S., and M. Gillespie, eds. *Technology, Innovation, and Southern Industrialization: From the Antebellum Era to the Computer Age*. Columbia: University of Missouri Press, 2008.

Deville, Winston. *Opelousas: The History of a French and Spanish Military Post in America, 1716–1803*. Cottonport, LA: Polyanthos, 1973.

Din, Gilbert C. *The Canary Islanders of Louisiana*. Baton Rouge: Louisiana State University Press, 1999.

———. *Francisco Bouligny: A Bourbon Soldier in Spanish Louisiana*. Baton Rouge: Louisiana State University Press, 1993.

———. *Spaniards, Planters, and Slaves: The Spanish Regulation of Slavery in Louisiana, 1763–1803*. College Station: Texas A&M University Press, 1999.

Ditto, Tanya Brady. *The Longest Street: A Story of Lafourche Parish and Grand Isle*. Baton Rouge, LA: Moran, 1980.

Dixon, Bill. *Last Days of Last Island: The Hurricane of 1856, Louisiana's First Great Storm*. Lafayette: University of Louisiana at Lafayette Press, 2009.

Donovan, F. R. *River Boats of America*. New York: Thomas Y. Crowell, 1966.

Dormon, James H., ed. *Creoles of Color of the Gulf South*. Knoxville: University of Tennessee Press, 1996.

———. *The People Called Cajuns: An Introduction to an Ethnohistory*. Lafayette: Center for Louisiana Studies, 1983.

Drago, H. S. *Steamboaters*. New York: Dodd, Mead, 1967.

Dronet, Curney J. *A Century of Acadian Culture: The Development of a Cajun Community: Erath (1899–1999)*. Erath, LA: Acadian Heritage and Culture Foundation, 2000.

Eakin, Sue. *Washington, Louisiana*. Bossier City, LA: Everett, 1988.

Eggler, W. A., A. Ekker, R. T. Gregg, E. Haden, A. Novak, R. P. Waldron, and H. B. Williams. *Louisiana Coastal Marsh Ecology*. Technical Report No. 14. Baton Rouge: Coastal Studies Institute, Louisiana State University, 1961.

Elliott, D. O. *The Improvement of the Lower Mississippi River for Flood Control and Navigation*. Vol. 1. Vicksburg, MS: US Army Corps of Engineers, Waterways Experiment Station, 1932.

———. *The Improvement of the Lower Mississippi River for Flood Control and Navigation*. Vol. 2. Vicksburg, MS: US Army Corps of Engineers, Waterways Experiment Station, 1932.

Falgoux, W. *Rise of the Cajun Mariners: The Race for Big Oil*. Ann Arbor, MI: Edward Brothers, 2007.

Fisk, H. N. *Geological Investigation of the Alluvial Valley of the Lower Mississippi River*. Vicksburg, MS: US Army Corps of Engineers, Mississippi River Commission, 1944.

———. *Geological Investigations of the Atchafalaya Basin and the Problem of Mississippi River Diversions*. Vicksburg, MS: US Army Corps of Engineers, Mississippi River Commission, 1952.

Fontenot, Mary Alice, and Paul B. Freeland. *Acadia Parish, Louisiana*. 2 vols. Baton Rouge, LA: Claitor's, 1976.

Frank, Arthur DeWitt. *The Development of the Federal Program of Flood Control on the Mississippi River*. New York: Columbia University Press, 1930.

Franks, K. A., and P. F. Lambert. *Early Louisiana and Arkansas Oil: A Photographic History 1901–1946*. College Station: Texas A&M University Press, 1982.

Gagliano, Sherwood M., and Johannes L. van Beek. *Environmental Base and Management Study: Atchafalaya Basin Louisiana*. Washington, DC: Environmental Protection Agency, 1975.

Gates, Paul W. *History of Public Land Law Development*. Washington, DC: Zenger Publishing, 1968.

Glasgow, Vaughn L. *A Social History of the American Alligator: The Earth Trembles with His Thunder*. New York: St. Martin's Press, 1991.

Gomez, Gay M. *The Louisiana Coast: Guide to an American Wetland*. College Station: Texas A&M University Press, 2008.

Griffin, Harry Lewis. *Attakapas Country: A History of Lafayette Parish, Louisiana*. New Orleans: Pelican Publishing, 1959.

———. *A Wetland Biography: Seasons on Louisiana's Chenier Plain*. Austin: University of Texas Press, 1998.

Guidry, Anita G. *La Pointe de l'Eglise: A History of Church Point, Louisiana, 1800–1973*. Lafayette, LA: Tribune Printing, 1973.

Guirard, Greg. *Atchafalaya Autumn*. St. Martinville, LA: privately printed, 1995.

———. *Cajun Families of the Atchafalaya: Their Ways and Words*. St. Martinville, LA: privately printed, 1989.

Guirard, Greg, and C. Ray Brassieur. *Inherit the Atchafalaya*. Lafayette: Center for Louisiana Studies, 2007.

Guirard, Greg, et al. *Psychotherapy for Cajuns: A Traditional Culture Struggles for Survival in a Crazy World*. St. Martinville, LA: privately printed, 2006.

Hackler, M. B. *Hurricane Rita and the New Normal: Modified Communication and New Traditions in Calcasieu and Cameron Parishes*. Jackson: University Press of Mississippi, 2010.

Hair, William Ivy. *Bourbonism and Agrarian Reform: Louisiana Politics, 1877–1900*. Baton Rouge: Louisiana State University Press, 1969.

Hallowell, Christopher. *Holding Back the Sea: The Struggle for America's Natural Legacy on the Gulf Coast*. New York: HarperCollins, 2001.

———. *People of the Bayou: Cajun Life in Lost America*. New York: Dutton, 1979.

Hanks, Amanda Sagrera. *Louisiana Paradise: The Chenières and Wetlands of Southwest Louisiana*. Lafayette: Center for Louisiana Studies, 1988.

Hatton, T. J., and J. G. Williamson. *The Age of Mass Migration*. New York: Oxford University Press, 1998.

Hebert, Kermit L. *The Flood Control Capabilities of the Atchafalaya Basin Floodway*. Rev. ed. Baton Rouge: Louisiana State University Press, 1973.

Heitmann, John A. *The Modernization of the Louisiana Sugar Industry, 1830–1910*. Baton Rouge: Louisiana State University Press, 1987.

Herring, H. W. *Some Facts about the Port of New Orleans*. New Orleans: Steeg Printing and Publishing, [ca. 1910].

Humphreys, Benjamin G. *Floods and Levees of the Mississippi River*. Washington, DC: Mississippi River Levee Association, 1914.

Hunter, L. C. *Steamboats on the Western Rivers*. Cambridge, MA: Harvard University Press, 1949.

Hyde, Samuel C. *A Fierce and Fractious Frontier: The Curious Development of Louisiana's Florida Parishes, 1699–2000*. Baton Rouge: Louisiana State University Press, 2004.

Hyde, Samuel C., et al. *The Manchac Swamp: Man-Made Disaster in Search of Resolution*. Hammond: Center for Southeast Louisiana Studies, 2006.

Jones, L. B., and G. R. Rice. *An Economic Base Study of Coastal Louisiana*. Baton Rouge: Center for Wetland Resources, Louisiana State University, 1972.

Kane, Harnett T. *The Bayous of Louisiana*. New York: William Morrow, 1943.

Kelley, M. J. *Nowhere to Hide: Defeat of the Sovereign Immunity Defense for Crimes of Genocide and the Trials of Slobodan Milosevic and Saddam Hussein*. New York: Peter Lang, 2005.

Kendall, John. *History of New Orleans*. Chicago: Lewis Publishing, 1922.

Kniffen, Fred B. *Louisiana: The Land and Its People*. Baton Rouge: Louisiana State University Press, 1968.

Kniffen, Fred B. *The Indians of Louisiana*. Gretna, LA: Pelican Publishing, 1965.

Kniffen, Fred B., H. F. Gregory, and G. A. Stokes. *The Historic Indian Tribes of Louisiana*. Baton Rouge: Louisiana State University Press, 1987.

Kolb, C. R., and J. R. van Lopik. *Geology of the Mississippi Deltaic Plain-Southeastern Louisiana*. Technical Report 2. Vicksburg, MS: US Army Corps of Engineers, Waterways Experiment Station, 1958.

Kondert, Reinhart. *Charles Frederick D'Arensbourg and the Germans of Colonial Louisiana*. Lafayette: Center for Louisiana Studies, 2008.

Kyvig, David E. *Daily Life in the United States, 1920–1939: Decades of Promise and Pain*. Westport, CT: Greenwood Publishing, 1920.

Laska, Shirley, and Andrew Puffer. *Coastlines of the Gulf of Mexico*. New York: American Society of Civil Engineers, 1993.

Laska, Shirley, et al. *Catastrophe in the Making: The Engineering of Katrina and the Disasters of Tomorrow*. Washington, DC: Island Press, 2009.

Lindstedt, D. M., L. L. Nunn, J. C. Holmes, and E. E. Willis. *History of Oil and Gas Development in Coastal Louisiana*. Resource Information Series No. 7. Baton Rouge: Louisiana Geological Survey, 1991.

Lockett, Samuel H. *Louisiana As It Is: A Geographical and Topographical Description of the State.* Baton Rouge: Louisiana State University Press, 1969.

Lockwood, C. C. *Atchafalaya: America's Largest River Basin Swamp.* Baton Rouge, LA: Beauregard Press, 1981.

Loos, J. L. *Oil on Stream: A History of Interstate Oil Pipe Line Company, 1909–1959*. Baton Rouge: Louisiana State University Press, 1959.

Martin, Michael S. *Historic Lafayette*. San Antonio, TX: Historic Publishing Network, 2007.

McCall, Edith. *Conquering the Rivers: Henry Miller Shreve and the Navigation of America's Inland Waterways*. Baton Rouge: Louisiana State University Press, 1984.

McGuire, T. *History of the Offshore Oil and Gas Industry in Southern Louisiana: Interim Report*. Vol. 2, *Bayou Lafourche—An Oral History of the Development of the Oil and Gas Industry*. OCS Study MMS 2004–050, 2004. US Dept. of the Interior, Minerals Management Service, Gulf of Mexico OCS Region, New Orleans.

McIntire, W. G. *Prehistoric Indian Settlements of the Changing Mississippi River Delta*. Coastal Studies Series No. 1. Baton Rouge: Louisiana State University Press, 1958.

McIntyre, T. *Seasons & Days: A Hunting Life.* Guilford, CT: Lyons Press, 2003.

McKenzie, L. S., III, and D. W. Davis. *Louisiana Gulf of Mexico Outer Continental Shelf Offshore Oil and Gas Activity Impacts.* Baton Rouge: Louisiana Mid-Continent Oil and Gas Association, 1994.

McNamara, Ed. *Cajun Racing: From the Bush Tracks to the Triple Crown*. New York: DRF Press, 2008.

Mehrländer, Andrea. *The Germans of Charleston, Richmond, and New Orleans during the Civil War Period, 1850–1870*. New York: Walter de Gruyter GmbH, 2011.

Michot, Stephen S., and John P. Doucet, eds. *The Lafourche Country II: The Heritage and Its Keepers.* Thibodaux, LA: Lafourche Heritage Society, 1996.

Morse, W.D. *The Birth of Jennings*. Jennings, LA: Jennings Chamber of Commerce, n.d.

Nardini, Louis Raphael. *No Man's Land: A History of El Camino Real*. New Orleans: Pelican Publishing, 1961.

Nostrand, R. L., and S. B. Hilliard, eds. *The American South*. Geoscience and Man, Vol. 25. Baton Rouge: Louisiana State University Press, 1988.

O'Neil, T. *The Fur Animals, the Alligator, and the Fur Industry in Louisiana*. Baton Rouge: Louisiana Department of Wildlife and Fisheries, 1977.

———. *The Muskrat in the Louisiana Coastal Marshes*. New Orleans: Louisiana Department of Wildlife and Fisheries, 1949.

Oubre, Elton J. *Vacherie: St. James Parish, Louisiana. History and Genealogy*. 2nd ed. Thibodaux, LA: Oubre's Books, [1986].

Owen, Donald E. *Geology of the Chenier Plain of Cameron Parish, Southwestern Louisiana*. Boulder, CO: Geological Society of America, 2008.

Pearson, C. E., G. J. Castille, D. W. Davis, T. E. Redard, and A. R. Saltus. *A History of Waterborne Commerce and Transportation within the U.S. Army Corps of Engineers New Orleans District and an Inventory of Known Underwater Cultural Resources*. Cultural Resource Series, Report Number: COELMN/PD 88/11. New Orleans: US Army Corps of Engineers, New Orleans District, 1989.

Peterson, R. W. *Giants on the River: A Story of Chemistry and the Industrial Development on the Lower Mississippi River Corridor*. Baton Rouge, LA: Homesite, 1999.

Post, Lauren C. *Cajun Sketches from the Prairies of Southwest Louisiana*. Baton Rouge: Louisiana State University Press, 1962.

Powell, Lawrence N. *The Accidental City: Improvising New Orleans*. Cambridge, MA: Harvard University Press, 2012.

Quick, H., and E. Quick. *Mississippi Steamboatin'*. New York: N. Henry Holt, 1927.
Reeves, W. D. *Louisiana: The Energy State*. San Antonio, TX: HPNbooks, 2013.
Reuss, Martin. *Designing the Bayous: The Control of Water in the Atchafalaya Basin, 1800–1995*. College Station: Texas A&M University Press, 2004.
Riffel, Judy, ed. *Iberville Parish History*. Baton Rouge, LA: Le Comité des Archives de la Louisiane, 1985.
Ronald, E. S. *Canals for a Nation: The Canal Era in the United States, 1790–1860*. Lexington: University of Kentucky Press, 1990.
Russell, R. J. *Glossary of Terms Used in Fluvial, Deltaic, and Coastal Morphology and Processes*. Coastal Studies Series No. 23. Baton Rouge: Louisiana State University Studies, Louisiana State University Press, 1969.
———. *River and Delta Morphology*. Coastal Studies Series No. 20. Baton Rouge: Louisiana State University Studies, Louisiana State University Press, 1967.
Russell, R. J., H. V. Howe, J. H. McGuirt, C. F. Hohm, W. Hadley Jr., F. B. Kniffen, and C. A. Brown. *Lower Mississippi River Delta: Reports on the Geology of Plaquemines and St. Bernard Parishes*. Geological Bulletin No. 8. New Orleans: Geological Survey, Louisiana Department of Conservation, 1936.
Schwartz, M., ed. . *Encyclopedia of Coastal Science*. Dordrecht, The Netherlands: Springer Netherlands, 2005.
Shlemon, R. J. *Development of the Atchafalaya Delta: Hydrologic and Geologic Studies of Coastal Louisiana*. Center for Wetlands Resources Report No. 8. Baton Rouge: Louisiana State University Press, 1972.
Sims, Julia. *Manchac Swamp: Louisiana's Undiscovered Wilderness*. Baton Rouge: Louisiana State University Press, 1996.
Sitterson, J. Carlyle. *Sugar Country: The Cane Sugar Industry in the South, 1753–1950*. Lexington: University of Kentucky Press, 1953.
Spitzer, Nicholas. *Louisiana Folklife: A Guide to the State*. Baton Rouge: Louisiana Folklife Program, 1985.
Spitzer, Nicholas, et al. *Mississippi Delta Ethnographic Overview*. Baton Rouge, LA: National Park Service, 1979.
Stirling, D. A. *A Bibliographic Guide to North American Industry: History, Health, and Hazardous Waste*. Lanham, MD: Scarecrow Press, 2009.
Swanton, John R. *The Indians of the Southeastern United States*. Washington, DC: Government Printing Office, 1946.
Taylor, G. R., and H. David. *The Transportation Revolution, 1815–1860*. New York: Rinehart, 1951.
Teurlings, William J. *One Mile an Hour: Priestly Memories*. New York: Exposition Press, 1959.
Theriot, Jason. *American Energy, Imperiled Coast: Oil and Gas Development in Louisiana's Wetlands*. Baton Rouge: Louisiana State University Press, 2014.
Trefethen, J. B. *The American Landscape: 1776–1976*. Washington, DC: Wildlife Management Institute, 1976.
Uzee, Philip D., ed. *The Lafourche Country: The People and the Land*. Lafayette: Center for Louisiana Studies, 1985.
Vermilion Historical Society. *History of Vermilion Parish, Louisiana*. [Abbeville, LA]: Vermilion Historical Society, 1983.
Vidrine, Malcolm F. *The Cajun Prairie: A Natural History*. Eunice: Louisiana State University–Eunice, 2010.
Widmer, M. L. *New Orleans 1900 to 1920*. Gretna, LA: Pelican Publishing, 2007.

Williams, T. Harry. *Huey Long*. New York: Random House Vintage Books, 1969.
Winters, R. K., G. B. Ward Jr., and I. F. Eldredge. *Louisiana Forest Resources and Industries*. Washington, DC: Government Printing Office, 1943.
Winther, O. O. *The Transportation Frontier: Trans-Mississippi West, 1865–1890*. Albuquerque: University of New Mexico Press, 1974.
Wood, S. W. *Live Oaking*. Annapolis, MD: Naval Institute Press, 1981.

Digital Media

Bellande, R. L. “Seafood.” http://biloxihistoricalsociety.org/node/36, *Biloxi Historical Society*, 2014.
Board of Engineers for Rivers and Harbors, War Department. [1924]. *The Port of New Orleans, La*. (Port Series, No. 5). Washington, DC: Government Printing Office. http://books.google.com/books?id=Bb81AQAAMAAJ&printsec=frontcover&dq=%22Po rt+series+no.+1%22+issues+4–6+Board+of+Engineers&hl=en&sa=X&ei=Z7P7UtaVJ- aH2gWj54DYAw&ved=0CDsQ6AEwAA#v=onepage&q=port%20of%20new%20orlean s&f=false.
“FM Global Urges Property Owners to Avoid Complacency Following U.S. Presidential Task Force Report on Hurricane Resiliency.” http://www.fmglobal.com/page.aspx?id=908212013.
Held, J. E. “The Canal Age.” *Archaeology Magazine Archive*. http://archive.archaeology.org/online/features/canal/.
“Industrial Revolution.” *The History Channel*. http://www.history.com/topics/industrial-revolution.
“The Age of Industrialization.” *Excellup*. http://www.excellup.com/classten/ssten/ageindustrialisation.aspx.

Dissertations

Knipmeyer, W. B. “Settlement Succession in Eastern French Louisiana.” PhD diss., Louisiana State University, 1956.
Mancil, E. “An Historical Geography of Industrial Cypress Lumbering in Louisiana.” PhD diss., Louisiana State University, 1972.

INDEX

A

B

W

Z